NOFA

Organic Principles and Practices Handbook Series
A Project of the Northeast Organic Farming Association

Crop Rotation and Cover Cropping

*Soil Resiliency and Health
on the Organic Farm*

Revised

SETH KROECK

Illustrated by Jocelyn Langer

CHELSEA GREEN PUBLISHING
WHITE RIVER JUNCTION, VERMONT

Originally published in 2004 as *Soil Resiliency and
Health: Crop Rotation and Cover Cropping on the
Organic Farm.*

Editorial Coordinator: Makenna Goodman
Project Manager: Bill Bokermann
Copy Editor: Cannon Labrie
Proofreader: Helen Walden
Indexer: Peggy Holloway
Designer: Peter Holm, Sterling Hill Productions

Printed in the United States of America
First Chelsea Green printing March, 2011
10 9 8 7 6 5 4 3   17 18 19 20

Library of Congress Cataloging-in-Publication Data
Kroeck, Seth.
  Crop rotation and cover cropping : soil resiliency and health on the organic farm / Seth Kroeck ;
illustrated by Jocelyn Langer.
      p. cm. -- (Organic principles and practices handbook series)
  "A Project of the Northeast Organic Farming Association."
  Originally published: Barre, MA : Northeast Organic Farming Association Interstate Council, 2004.
  Includes bibliographical references and index.
  ISBN 978-1-60358-345-9
  1. Crop rotation--United States. 2. Cover crops--United States. 3. Organic farming--United States.
I. Langer, Jocelyn. II. Title. III. Title: Soil resiliency and health on the organic farm. IV. Series: Organic
principles and practices handbook series.

  S603.K76 2011
  631.4'52--dc22

                              2010051157

Chelsea Green Publishing
85 North Main Street, Suite 120
White River Junction, VT 05001
(802) 295-6300
www.chelseagreen.com

CONTENTS

Best Practices for Farmers and Gardeners

The NOFA handbook series is designed to give a comprehensive view of key farming practices from the organic perspective. The content is geared to serious farmers, gardeners, and homesteaders and those looking to make the transition to organic practices.

Many readers may have arrived at their own best methods to suit their situations of place and pocketbook. These handbooks may help practitioners review and reconsider their concepts and practices in light of holistic biological realities, classic works, and recent research.

Organic agriculture has deep roots and a complex paradigm that stands in bold contrast to the industrialized conventional agriculture that is dominant today. It's critical that organic farming get a fair hearing in the public arena––and that farmers have access not only to the real dirt on organic methods and practices but also to the concepts behind them.

About This Series

The Northeast Organic Farming Association (NOFA) is one of the oldest organic agriculture organizations in the country, dedicated to organic food production and a safer, healthier environment. NOFA has independent chapters in Connecticut, Massachusetts, New Hampshire, New Jersey, New York, Rhode Island, and Vermont.

This handbook series began with a gift to NOFA/Mass and continues under the NOFA Interstate Council with support from NOFA/Mass and a generous grant from Sustainable Agriculture Research and Education (SARE). The project has utilized the expertise of NOFA members and other organic farmers and educators in the Northeast as writers and reviewers. Help also came from the Pennsylvania Association for Sustainable Agriculture and from the Maine Organic Farmers and Gardeners Association.

Jocelyn Langer illustrated the series, and Jonathan von Ranson edited it and coordinated the project. The Manuals Project Committee included Bill Duesing, Steve Gilman, Elizabeth Henderson, Julie Rawson, and Jonathan von Ranson. The committee thanks SARE and the wonderful farmers and educators whose willing commitment it represents.

Special Note:

Organic Farming techniques are valuable ideas and tools to consider in the pursuit of profitable

Acknowledgments

The author would like to acknowledge the extensive research and trials done by many talented authors and farmers in the Northeast. A handful of farmers contributed directly to this manual through interviews and correspondence and their acknowledgment here is in thanks for their generous advice during the busy summer months. These farmers include Dan Kaplan, Brookfield Farm; Julie Rawson, Many Hands Farm; Jean-Paul Courtens, Roxbury Farm; Eileen Droescher, Ol' Turtle Farm; Jim Crawford, New Morning Farm; and Dave Colson, New Leaf Farm.

one

Introduction

What is a successful organic farm? Is it one that meets the bottom line? Is it one with relatively weed-free fields, or maybe a steady increase in soil fertility? Does it produce flawless crops, unblemished by disease and pests? Most growers would say yes to each of these questions and maybe add another two or three benchmarks of their own. With a little thinking the list can get pretty long, and the thought of answering each question individually for your farm might be overwhelming. The trick to keeping your head on straight and getting through the day's work is tying many of these objectives together with one tool: a rotation plan. The goal of this handbook is to outline the benefits and describe how to create good crop-rotation practices, including the use of cover crops, for northeast organic farms. *Hopefully some of these practices can be applied to other parts of the USA.*

Eight years ago I began working on a small organic farm in California as an apprentice. Fueled by enthusiasm more than knowledge, I moved to the Northeast the following season and became a hired hand on a farm in Massachusetts. Working from early spring until the ground began to freeze in the fall, I was intrigued by the quick succession of crops, awakening slowly, peaking, then withering gradually, all in the space of six or seven months. This cycle of short, intense growing seasons followed by cold winters seemed to be tied together only by the rotation plan we followed and the focus of the farmers I was working for, who kept one eye on the future.

I began to study different rotations while working on a succession of northeast farms in the following seasons. Every farmer had a different perspective on what made a farm successful as well as a particular aspect of the operation that he did with special commitment. I had the best

Hairy vetch.

experience and learned the most from the farmers who were committed to organizing their operations from top to bottom and understanding every detail to the point that they could answer almost any question. Invariably these were the growers who had also devised crop-rotation systems that addressed fertility goals for the farm, maximized the health and diversity of their fields, and found a manageable compromise between a strict rotation and making a profit.

As a manager myself now, the skills shared by many northeast growers have helped me get started. Creating a rotation plan for my farm was central to allowing me to see beyond the season's work to larger goals of steward-ship and growth. When this writing opportunity came my way, I saw a chance to dig deeper into northeast rotations, talking to more farmers and finding current research that could add to my knowledge and help create a resource for other growers. As the handbook has evolved I have found three statements that distill the ideas it contains and just might persuade you to take the plunge and commit your farm to a rotational system:

1. The foundation of organic management is the creation of a healthy farm system that builds soil fertility and prevents the proliferation of pests, weeds, and disease.
2. Well-timed rotations between cash and cover crops effi-ciently address both of these goals when instituted over the course of many seasons.
3. Finding the compromise between perfect horticultural practice and the logistical flexibility needed to be profit-able is the mark of a successful northeast rotation.

In the pages that follow you can learn some of the thinking behind—and the actual nuts and bolts of—laying out a rotation for your farm. Addressed are many of the possible variables in forming the rotation as well as where to find room for compromise between the "perfect horti-culture" model and day-to-day logistics. Tables with helpful information about crop families, cover crops and their uses, equipment costs, sample rotations, and so on, are interspersed throughout the text. And interviews are included with two growers, accompanied by samples of the rotations they use on their farms.

To be sure, creating a crop rotation for a highly diversified northeast organic farm is challenging. It involves hard thinking and serious compromise between your immediate financial goals and the long-term goals of the farm. The reward, however, is a healthier operation, with benefits ranging from your soil's health to your bottom line. A crop-rotation plan for your farm is no silver bullet, but over the course of several seasons it can, quite simply, improve the quality of the crops you grow.

two

Historical Roots

Farmers in the Northeast have been rotating crops one way or another since Native Americans brought cultivation techniques from the Midwest to the silt and upland soils along our region's rivers. While crops, techniques, and the cultures have changed over the centuries, each era has reaffirmed the importance of rotation.

Native Americans considered the land along the rivers the most productive (as do most growers today), gaining fertility from the regular flooding that brought additions of rich silt. Areas along these waterways were easily cleared for growing a long list of crops including flint corn, beans, squash, pumpkins, watermelons, tobacco, and Jerusalem artichokes, which were sown together in the fields.

Rotation systems for the first nonnative farmers in the Northeast involved several intense seasons growing the Indians' version of "cash crops" and long fallow periods. The fallows allowed weeds to reclaim fields and stabilize fertility while minimizing erosion with quick soil-covering growth. In fallow years, game animals crossed the open, weed-covered fields as they moved toward the waterways, allowing the plots to double as hunting grounds. Once the soils were revitalized by weed growth, the native farmers would burn off the surface vegetation, the ash releasing potassium and lime that conditioned the soil. With shallow hand tools that minimized the disruption of the soil, they would till in the roots of the weeds in preparation for sowing. The planting of flint corn, the heaviest feeder among their crops, coincided with the run of herring in the rivers along the fields. Scooped up and added to the hills beneath the seed, these fish provided a rich nitrogen source for the corn and lower-story crops, like squash, that were commonly sown.

The first European settlers learned and followed these techniques (buying large quantities of native produce in order to survive during the process) until the first plow and livestock arrived. At this point, European methods—for example, cropped land rotated with livestock pasture for a set period—became the standard, and large tracts of northeast forest fell to the axe and plow.

It was common for the leaders of early European colonies to prescribe rotations for their member farmers, specifying the succession to be grown and the number of years land should rest as pasture, with both dependent on the quality of the soils. Forage-sown pastures and grain production on cropland provided the organic matter of today's cover-cropping methods. After the Revolutionary War, many of these systems continued. Many townships designated common pastureland for animals during the time they were rotated off their owners' cropland.

Larger-scale agriculture, and the farm-industrialization boom that originated in the late 1800s on the vast prairies of the Midwest, began to influence northeast farmers, pulling them away from longer, more diverse rotations. The proliferation of machinery catapulted small northeast farms into an economic system that was built on scale and specialization. Diverse rotations designed for sustainability and followed for generations gave way to two- and three-crop systems that evolved around commodity markets and large-scale distribution. Grain production moved to the

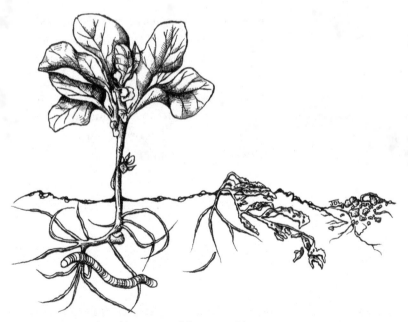

Buckwheat in growth and in stages of decomposition.

Midwest where rich soils and large acreage could produce it more cheaply, and the completion in 1825 of the Erie Canal provided the method of transport to markets in the East. Dairy became the crop of choice in the Northeast, with farmers supplying a growing urban population with what still needed to be produced locally. Vegetables were still also grown locally, but a reliable rail system and the advent of trucking began to bring perishables from outside the region in increasing supply. Vegetable growers, too, gradually adopted the fashion of specialization and produced fewer and fewer crops from larger plantings.

As the middle of the twentieth century arrived, the advent of Green Revolution ideas further eroded the rotation systems of the past with the introduction of high-yield cereal grains, NPK (nitrogen/phosphorus/potassium) fertilization, and nozzle-administered controls for weeds and pests.

In the last thirty years, northeast farmers have begun returning to more diverse, soil-enriching rotations. Years of input-focused farming had taken its toll on the structure of our most fertile land, prompting a rebirth of old methods and rotations fine-tuned by new scientific understanding and techniques. In the past thirty years, the emergence of organic agriculture has played a leading role in revitalizing sustainable agriculture in the Northeast.

The best rotation plans on today's northeast farms echo an earlier diversity and flexibility while bringing measurable improvements in organic matter and nitrogen availability. These successful rotations promote fertility in the long-range, employing annual soil analysis and intensive cover cropping, but are also flexible in the short term to allow for seasonal tailoring of an individual farm's production plan.

three

Thinking Beyond This Season's Cash Crop

Why Make the Effort?

The reason to institute a rotation plan can be summed up in one thought: to create a healthy farm system. And this, beyond any other focus, will increase a farm's productivity over the course of many seasons. The quality of a farm operation is tough to quantify and, beyond the profit-and-loss statement, there are few ways to assess it year to year. The addition of rotation and cover cropping doesn't create a way to quantify it, but there are five aspects of your farm system that are likely to show improvement if monitored over several seasons: overall efficiency, soil fertility, and the reduction of disease, weeds, and insect pests. Knowing that a good rotation plan can benefit your farm in all these ways can be a great motivator toward the decision to create one for your farm.

Efficiency

Creating a good rotation plan can be a first step in streamlining all the everyday tasks that make up the production, harvest, and marketing of a farm's crops. Like an outline for a writing project or a blueprint for a new building, a rotation plan can be the structure that holds together several complex ideas and ensures that they work well together. It can also streamline and focus management decisions during the heat of the season—so there's that little bit of extra brainpower available for making all the other aspects of the farm run more efficiently. For example, a good rotation map can remind you that there are three half-acre units to spread compost on before the fall cover crop of oats/peas can be sown. Knowing the exact size of these units allows quick calculations of how much you'll need, how long it will take to spread, and if you have enough oats and peas to sow. On top of this benefit for the farmer, being able to communicate the farm's big picture on paper to an assistant or even the whole labor crew can clarify things for hired help and make the experience more interesting for them as well.

Not just daily productivity, but also the ecological health of the farm benefits over time from a rotation. Less time and money spent each year on amendments for fertility, pest and disease controls, and beating back weeds calculate into savings at the end of the year. Pennsylvania growers Anne and Eric Nordell spell out the savings they've seen in just one of their crops since beginning their rotation: "It took us five years to develop the . . . system, which has kept the in-row hand weeding of onions well below 15 hours per acre."[1] The forethought and work can end up repaying real savings that grow with the success of your efforts.

Diseases

A good rotation will decrease soil-borne disease and pest outbreaks. These two pressures on the organic farm are often the primary motivation for growers to create a rotation, as the damage of one outbreak can decimate several cash crops in just a few days.

Most diseases that affect cash crops reflect an imbalance in soil ecology. Microorganisms associated with disease are present in healthy soils to some degree, but they exist in a delicate balance alongside microscopic relatives that perform beneficial, even vital, tasks for the crops. It's when populations fall out of balance that disease becomes a problem.

It's when a crop (or family of crops) grows in the same location season after season that disease populations can increase. Plant families, and plants with similar rooting characteristics, follow comparable patterns in the nutrients they gather and the microorganisms they foster. After several seasons supporting the same crop, soils can become depleted of specific nutrients necessary to the development of that crop, which, in turn, makes the crop more susceptible to disease (and pests, for that matter). The populations of pathogens of a specific crop will explode as they prey on it while other microorganisms in the soil ecosystem will suffer or even disappear. The imbalance can affect succeeding plantings.

Intensive tillage and cultivation can encourage soil disease by disrupting soil ecology. Every time soil's natural layering is destroyed, water and air-holding aggregation break down, organic matter is wasted, and compaction is encouraged. These disruptions reduce the exchange of air and water though the active layers of the soil. This starves or suffocates large parts of the beneficial microscopic population. What's left is an environment that, from nature's point of view, needs to be brought back into balance, and that the farmer experiences as crop disease.

Insect Pests

Crop plants thrive on the diversification of soil culture and nutrients that rotation provides, making them healthier and consequently better able to resist pest damage in the first place. As with diseases, most pest populations explode when offered the same crop in the same location season after season. They winter over in the field or in the surrounding environment and emerge season after season to take advantage of their favorite crop, considerately supplied by the farmer. Moving the crop to another field on the farm or even another area in the same field provides a long, stressful trip for some pests that emerge in spring and summer, reducing the initial population and that of generations to come.

The diversity created when cash and cover crops are sown in small units, in a checkerboard pattern within a single field, can help control pest populations. Fields sown with a diversity of crops provide varied habitat for numerous insect populations, which keeps outbreaks of a single species in check. Beneficial insect species control pest species through predation or enhance crop yields as pollinators. They also provide the first line of defense against seasonal pests that don't overwinter, but blow in every season on weather patterns from the South and Midwest.

Weed Suppression

Weed control on the organic farm is a major concern and can overwhelm many growers, turning their farms into jungles that strangle cash crops and perpetuate the less-than-positive "organic weed-farmer" image. Tackled solely with the hoe or tractor, weeds can dog a grower for the life of the farm. The addition of a rotation that includes variations in tillage depth and timing, cover crops, and careful use of bare-fallow periods can visibly reduce weed populations over the course of just a few seasons. In fact, many growers begin using cover crops specifically for weed control.[2]

Like soil diseases and pests, weeds are experts at finding a niche in

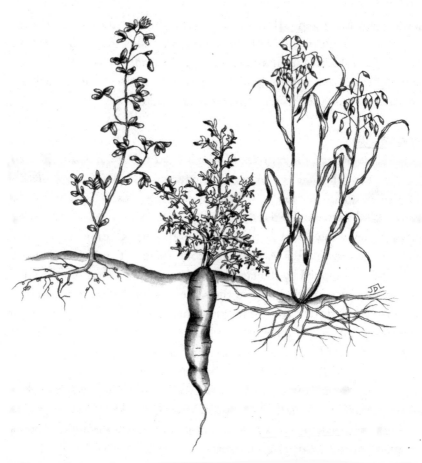

Each crop influences the soil ecology differently. A legume cover crop like alfalfa (*left*) will add nitrogen. Tap-rooted crops like carrots (*middle*) loosen soil compaction and transport minerals from the subsoil to the topsoil. Grasses (*right*) repair soil structure.

an unbalanced ecosystem. Most cash crops are sown, cultivated, and harvested at the same times every year, and weed populations find their place in this timing to mature, set seed, and return stronger the next year. On some farms this phenomenon can be so clear that it is possible to diagnose problems with rotations by the strength of particular weed species. Organic growers who have taken over conventional corn fields, not rotated for a generation, know this story too well with bindweed and grasses that are almost unstoppable in the first years of transition.

Weed suppression on a farm takes a lot of planning and understanding on how to fight these monsters on a yearly basis. One must completely know every aspect of this enemy.

When cash and cover crops are rotated, the change of sowing, culti-*Insects and weeds are the No. 1 enemies* vation, and tillage times on a given field disrupts the life cycle of weed species—annual or perennial—preventing seed set and bringing root systems to the surface. A cover crop sown after, or intersown with, a maturing cash crop can also reduce weed pressure by shading out and smothering weeds. For problem areas of the farm, or in response to changing weed pressure, a grower might choose a specific cover or cash crop based on its competitive qualities—e.g., its strong canopy, its need for frequent cultivation, how early it is put into or taken out of the field, and so on. Bare fallowing, or leaving fields open and allowing weeds to germinate and emerge before harrowing them in ("stale seedbed"), is also effective in controlling strong weed populations. If done in short periods of two to four weeks at a time, it is possible to minimize the loss of organic matter and soil structure that are side effects of tillage and keeping the soil uncovered. Combining the smother qualities of cover crops with short two-to-four-week fallow periods over the course of a full season can further break their life cycle and reduce the overall weed seed bank. No matter how good the rotation, it will take a few seasons for the full effect on weed populations to show.

Soil Fertility

Soil fertility is mysterious. The balance of thousands of living inhabitants with organic and mineral components, along with air and water, creates a living cycle of production, consumption, and decomposition that defines soil's health. Rather than practice green-revolution science that reduces the soil ecosystem to its NPK (nitrogen, phosphorus, and potassium) needs, organic growers choose to protect its complexity for success in the long term. Stewardship of soil cycles and fertility is difficult on the farm without a good rotation.

Only a resilient soil ecosystem reliably provides nutrients for crops. Rotating different families of cash and cover crops (as well as composted animal manure and mineral amendments) into the biology of the soil over the course of many seasons allows the grower to play a role in shaping its health and thus fertility. A given crop may leave behind nitrogen, improved aggregation, and increased organic matter for the next crop. Tap-rooted crops open up compacted soils and bring minerals from the subsoil to the topsoil. Small-seeded grasses repair soil structure with dense root systems. Fast-growing grains capture soluble nutrients for the crops that follow.

Be careful when it comes to tillage and what is trying to be accomplished. It appears, at this time that 2-3 inch

Tillage

A rotation plan can involve variations in the types of tillage, which can benefit fertility by preserving and improving soil biology and structure. Many northeast farmers believe that alternating the plow (or other deep soil-turning tillage tools like some spaders and rotovators) with shallow cultivation is as vital to soil and crop health as rotating plant families. Varying the depth and the degree to which soils are turned over and aerated can influence both the long-term health of the soil and the short-term vitality of the season's crop.

Farmers have traditionally used the plow to turn under a crop before the planting of the next one, thus aerating soils deeply to allow crop roots to penetrate easily. Working at a good depth, the plow maximizes the incorporation of lots of living or dead residue and does an exceptional initial job of preparing a fine soil surface for easy seeding and cultivation. The disadvantage to tilling deeply is that it can be tough on your soils. It mixes the soil to whatever depth the tool is set, commonly bringing up some subsoil and weed seeds from below and putting the most fertile, biologically active topsoil down, away from shallow-rooting cash crops. Valuable moisture comes to the surface where it can evaporate. Air gets mixed in at all levels, stimulating microbes that then mineralize valuable organic matter too quickly for crop plants to utilize. This in turn damages the soil's structure and its ability to hold nutrients and water. Deep tillage also exposes more soil to wind and water erosion by loosening it to greater depths.

Concern for avoiding these pitfalls has caused some farmers to alternate the plow with shallow tillage techniques that work only the top few inches of the soil. These growers have found that a disc harrow, a sweep-type field cultivator, a modified shallow-depth plow, or a rotovator will incorporate crop residues well at shallower depths. The process becomes easier if the residue is first chopped closely with a flail-type mower and, if green, allowed to wilt down. Working these residues into just the surface preserves soil moisture, keeps old weed seed from being exposed, maintains the proper soil horizon (topsoil on top, subsoil below), and folds in organic matter. Shallow tillage leaves roots in place to decompose, enhancing deeper soil structure. It can also help warm surface soil by adding air, which insulates it incrementally from the colder ground beneath and encourages crop residues to decompose more quickly.

There are implements that can provide the deep aeration qualities of the plow while minimizing soil damage. By working soil at 12- to 18-inch depths without turning it over, chisel plows, subsoilers/rippers, and reciprocating spaders (their spades work in an up and down motion) can be effective soil-opening tools that are gentle with soil structure.

Alternating shallow with deep tillage allows a farmer to plow with less worry of damaging the soil, while still gaining the deep soil opening and residue-incorporation benefits.

As the plan below shows, Eric and Anne Nordell have built deep and shallow tillage into their rotation practice. They match tillage tools and depth with the incorporation tasks for which the tools are best suited. Deep plowing is reserved for fallow years, when the weed seeds and soil disruption it can bring will not affect cash crops.

Year 1: deep plow rye / bare fallow /sow oats & peas

Year 2: early harrow oats & peas residue / sow early onions / sow clover

Year 3: deep plow clover / bare fallow / sow rye & vetch

Year 4: skim plow rye & vetch / sow late cash crops / sow rye

Year 5: deep plow rye / bare fallow / sow oats & peas

Year 6: early harrow oats & peas residue / early crops / sow clover[3]

No-till. Researchers in the Northeast are currently working on farm-scale systems of not tilling at all, and they are gaining steam. These no-till rotations use cover crops as "killed mulch" for transplanted or direct-seeded crops that are planted right into them. This keeps soil covered, reducing weed pressure, preserving moisture, and virtually eliminating soil disturbance from tillage and cultivation. Unfortunately, this practice is limited for organic growers as systems for many crops still rely on herbicides for management. Developing implements that will knock down and kill many types of cover crops reliably is the next goal. In the years to come rotations and soil health may benefit significantly from these innovations. (Pennsylvania farmer Steve Groff is leading the way in no-till vegetables. His Web site, listed in the "Resources" section, is a great introduction to these ideas).

Weed Control

Each crop is unique in its susceptibility to weed pressure. Some are vigorous and compete well with weeds, while others can be quickly overwhelmed. Crops also vary in the ease with which they can be kept weed-clean. By alternating crops that are good competitors with those that are not—or those that are easily cultivated with those that are difficult—it is possible to reduce weed pressure over time. Preceding a difficult-to-cultivate crop like winter squash with one that is easy to keep clean, like potatoes, can help reduce weed pressure on the squash. Following a heavily-cultivated crop like onions with one that requires little soil disturbance, like sod for pasture, can help maintain soil structure and maximize weed suppression.

Animal Factors

Manure, whether left by animals on rotated pasture or applied as compost, contributes greatly to fertility management in a rotation. Alongside crops and microorganisms, livestock are important contributors to the soil ecosystem, filling a unique macro-role in the cycle of energy capture, consumption, and decomposition that defines living soil.[4] The digestive tracts of animals process plant material in a unique way, releasing nitrogen, phosphorus, and potassium in a form that's immediately available to plants and microorganisms. In addition, manure inoculates soil with diverse, numerous decomposer organisms from the animals' digestive tracts that improve the soil's natural ability to break down organic matter. Applied composted manure and/or pasture-grazed livestock contribute to long-term soil health.

Cash and Cover Crops

Every crop in a rotation will have some interaction with the soil's fertility. Some of these effects may become obvious during a single season, but many will take several years to observe. For example, deep-rooted cash and cover crops like beets or alfalfa pull minerals up from the subsoil, making them available to subsequent sowings of shallow-rooted crops. Their deep rooting action also helps open compacted ground for better drainage. Crops with dense root systems like annual ryegrass gather soluble nutrients, leaving them in the soil for the next crop as their root systems decay. Cover crops in particular can act as gatherers of soluble nutrients. Fast-growing grains and densely-rooted grasses gather large amounts of nitrogen, storing it in their leaves until tilled in, then releasing this nutrient for the following crop. Sowing a dense cover immediately after spreading compost can be the most efficient way to save compost nutrients for metabolizing by the soil. Unlike many cash crops, covers are able to quickly develop a hungry root system and monopolize nutrient-gathering. They also establish themselves over the whole area instead of only in the rows/beds where cash crops are sown. Finally, planting cover after manure compost is spread constructively utilizes the time between the application of animal waste on the field and the harvest of a food crop, a strategy that can help certified growers comply with the current USDA rules for handling compost.

Cash crops are needed to keep the farm or ranch afloat. You really need to have a program in place that allows you have positive cash flow and soil build up from a good cover crop. This cover crop rotation is really an intensive application of plant knowledge. - Really need to know your "stuff" to make cover crops a profitable enitity for the overall situation.

four

What Is a Good Rotation?

Northeast organic growers farm smaller, more diverse acreage than most farmers around the country and use many different criteria to create rotational systems for their individual farms. Pulling together all the variables that influence a cropping system and integrating them into a rotation that can be used year after year requires a comprehensive look at the whole farm and the ability to make educated compromises between production needs and long-term soil health. (See Henderson and North, *Whole-Farm Planning*—a handbook in this series that examines the "whole" you have to work with and helps you identify your goals as a farmer.)

This chapter identifies many of the elements a northeast grower will use to build a rotation, then discusses different ways of mapping fields, grouping crops, introducing cover crops and tillage, and finally, ties all these pieces together into a rotation. Like farming in general, there is no $a + b = c$ equation to building a successful rotation. The goal here is to identify the major pieces, show a few ways to fit them together, and illuminate the process as a whole so you can adapt the ideas to your farm.

This outline diagrams the process:

1. Identify central reasons to rotate crops on your farm.
2. Create a map of your farm.
    - Note important features (roads, irrigation, variations in soil fertility and wet areas, etc.)
    - Divide the farm into "units" of the same size that define plots that crops will rotate through.
3. Group your crops for rotation.
    - List all of the factors that could influence certain crops to be grouped with others in the field (time of seeding/planting, time of harvest, acreage needed, plant family, nutrient needs, etc.).

- Categorize all of your crops by each factor and include cover crops.

4. Add crops to your map.

- Put crops (including covers) into rotation units, grouping them together according to priority, defined by the lists of important factors.
- Consider features of the farm when putting each crop into the map (irrigation, wet spots, access for harvest/cultivation, etc.).
- Double-check details of planting dates and acreage needed for each crop to ensure everything fits in each unit.
- Add details of management for each unit (tillage type and dates, mowing, seeding covers, etc.).

Create a Map of Your Farm

Details of the Farm

Map out the tillable acreage on which you would like to rotate crops. The map should be a clear, simple representation of your farm that notes landmarks, roads, irrigation, access, and variations in soil. It is especially important to note which sections are wet in the spring, overly dry in midsummer, less fertile, and under heavy weed pressure, as these factors will affect your plans, e.g., what crops can be rotated into those areas or when. If the farm sprawls over several fields, try to contain each field on one page and keep the pages somewhat in scale, then tack the pages together on a wall to display the whole farm at once.

Units for Rotation

Once a map is complete, the process of creating units that define the plots in the rotation can begin. To make the system work smoothly and flexibly, it's necessary to create units that are roughly the same size. (It's also helpful to make the size of units some fraction of an acre, since seeding and amendment rates are usually given in pounds per acre). Decide how

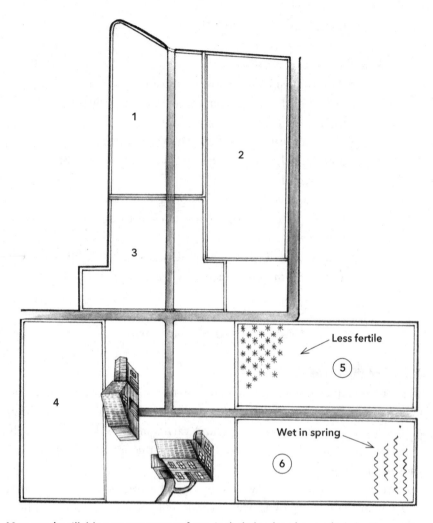

Map out the tillable acreage on your farm. Include landmarks, roads, irrigation, variations in soil, and so on. Note areas that are wet in spring, dry in midsummer, or less fertile, and such things as where weed pressure is especially heavy.

much total acreage you have to work with in your rotation. Making your beds and rows a consistent width will make it easy to calculate your total tilled acreage and the acreage assigned to each individual crop. In new plots that have not been cropped before, walk off measurements of width and length that can be translated into x number of beds/rows times y lengths. For example:

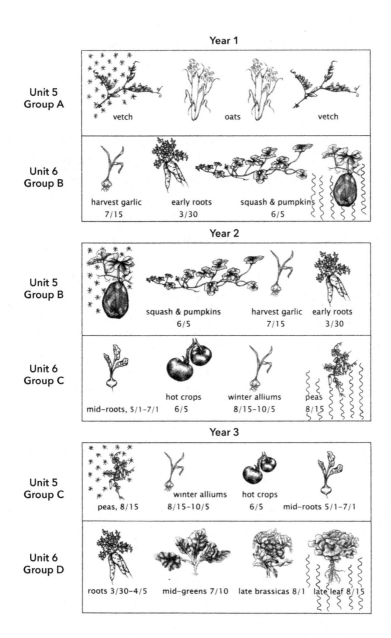

A sample three-year rotation for units 5 and 6 (on the farm map in previous figure) would take into consideration their wet and less-fertile areas. Vetch, pumpkins/squash, and peas grow well in-less fertile areas because of their lower nutrient needs. Squash, peas, lettuce, and other crops planted after June 1 can do well in an area that was wet earlier in the spring.

10 beds @ 42 in. (3.5 ft.) wide × 250 ft. long = 8,750 sq. ft.
   1 acre = 43,560 sq. ft.
   43,560 divided by 8,750 = 5
Those 10 beds equal 1/5 acre

Each unit should be a functional size without creating too many or too few units in relation to the overall number of acres on your farm. Many growers I know use anywhere from 4 to 20 units. A few large units can provide a lot of flexibility when laying out many crops that require different acreage, while many small units can help create a more structured rotation for complex successions of plantings. A common unit size for a 5- to 10-acre farm might be ¼ to ½ acre. Many growers devise unit sizes around features of the farm already in place like road systems or irrigation layouts. Reconfiguring the layout of these features to fit your new rotation units can also be a good investment in productivity by allowing the laying of irrigation or access roads between units, or establishing permanent "biostrip" habitat for beneficial insects. Variations in soil fertility, or seasonal wetness or drought, should also be considered at this point. If possible, these areas should constitute individual units so that crops that can tolerate the conditions can be grown there and improvement strategies included in the rotation.

With units created and marked on the map, make multiple copies of each field and document the past 4–5 years of cropping on the farm in as much detail as possible, each year on a different set of map pages. If fields have been in sod, try to tease out what species are there, including weed populations, as this will influence what cash crops you may want to plant there in the future.

Grouping Crops

Now that you have a map showing your new units, think of it as a blank canvas and set it aside to do some sketchwork before you begin painting.

All of your crops (cover and pasture included) can be loosely grouped in several ways—based on plant family, time of seeding and harvest, acreage needed, nutrient needs, and many other variables that may be unique to your farm.

When grouping crops for rotation, consider:

- time of seeding/planting
- time of harvest
- acreage needed
- plant family
- nutrient needs
- tillage needs
- wetness/drought tolerance
- ability to compete with weeds
- cultivation needs
- irrigation needs
- pest pressure
- harvest access
- benefit from preceding crop
- frost tolerance

Commonly, farmers are advised to use plant families as the most important factor in grouping crops, and this association is indeed vital as it relates to fertility and disease and pest control. However, most growers find groupings based first on timing—of field prep, seeding, and harvest—to be more practical. By detailed mapping or the creation of permanent bed or row systems, a farmer can set up rotations based on plant family *within individual units*, creating a compromise between these two grouping factors. By making lists of your crops and grouping them different ways, you will be able to find the factors that are most important to the short-term productivity of your farm and its long-term soil health. As Jean-Paul Courtens of Roxbury Farm in Kinderhook, New York, puts it: "I sow 50 percent of my ground to two years of cover crop, so—as long as I don't plant the same cash crop families in the same place in successive years—I can be guaranteed a four-year break between them."

Some Groupings of Typical Crops
The following lists are of typical vegetable and cover crops grouped according to some common factors in rotation. Compatible cover crops

are shown in italics with cash crops to suggest their inclusion—either interplanted or alongside the cash crop. For ease of reference, all of the characteristics of cash crops listed in this section are spelled out in table 1 (pages 24–25).

Field Seeding/Planting Date

Early (March–May): onions, beets, chard, spinach, lettuce, endive, broccoli, cabbage, kale, peas, potatoes, carrots, parsnips, parsley, strawberry, *cereal rye, ryegrass, fescues, barley, oats, clover, vetch*

Midseason (June-July): beets, chard, lettuce, endive, cabbage, kale, turnips, cucumbers, squash, melons, pumpkins, corn, beans, peas, tomatoes, peppers, eggplant, carrots, parsley, celery/celeriac, *cereal rye, ryegrass, millet, sorghum-Sudan grass, fescues, barley, oats, clover, vetch*

Late season (August–November): onions, garlic, chard, spinach, lettuce, endive, broccoli, cabbage, kale, turnips, peas, *cereal rye, ryegrass, millet, sorghum-Sudan grass, fescues, barley, oats, clover, vetch*

Family

Amaryllidaceae (allium)	onions, garlic
Chenopodiaceae (beet)	beets, chard, spinach
Compositae (lettuce)	lettuce, endive
Cruciferae (crucifer)	broccoli, cabbage, kale, turnips
Cucurbitaceae (cucurbit)	cucumbers, squash, melons, pumpkins
Gramineae (grass)	corn, *cereal rye, ryegrass, millet, sorghum-Sudan grass, fescue mixtures, barley, oats*
Leguminosae (legume)	beans, peas, *clover, vetch*
Rosaceae (rose)	strawberries
Solanaceae (nightshade)	tomatoes, peppers, eggplant, potatoes
Umbelliferae (carrot)	carrots, parsnips, parsley, celery/celeriac

Length of Time in Field

60 days or less: chard, spinach, lettuce, endive, *ryegrass, oats, millet*

90 days or less: onions, garlic (planted previous fall), beets, carrots, broccoli, cabbage, kale, turnips, cucumbers, melons, corn, beans, peas, *cereal rye, ryegrass, millet, sorghum-Sudan grass, fescues, barley, oats, clover, vetch*

120 days or less: beets, carrots, squash, pumpkins, tomatoes, peppers, eggplant, potato, leeks, parsnip, parsley, celery/celeriac, *cereal rye, ryegrass, millet, sorghum-Sudan grass, fescues, barley, oats, clover, vetch*

2-year perennial: strawberries, Italian ryegrass, sweet clover

Typical Planting Sizes

Large (10%+ of cropped land): squash, pumpkins, corn, *cereal rye, ryegrass, millet, sorghum-Sudan grass, fescue mixtures, barley, oats, clover, vetch*

Medium (2%–5% of cropped land): lettuce, broccoli, cabbage, kale, turnips, tomatoes, peppers, eggplant, potatoes, strawberries, *cereal rye, ryegrass, millet, sorghum-Sudan grass, fescue mixtures, barley, oats, clover, vetch*

Small (1%–2% of cropped land): onions, garlic, beets, chard, spinach, endive, cucumbers, melons, beans, peas, carrots, parsnips, parsley, celery/celeriac, *cereal rye, ryegrass, millet, sorghum-Sudan grass, fescue mixtures, barley, oats, clover, vetch*

Nutrient Needs

High: onions, garlic, broccoli, cabbage, kale, turnips, corn, *ryegrass, millet*

Medium: beets, chard, spinach, lettuce, tomatoes, peppers, eggplant, potatoes, carrots, parsnips, parsley, celery/celeriac, *oats, sorghum-Sudan grass, fescues*

Low: cucumbers, squash, melons, pumpkins, beans, peas, *cereal rye, barley, clover, vetch*

Plant Part Harvested

Leaf: cabbage, celery, chard, endive, kale, lettuce, parsley, spinach

Fruit/flower: beans, broccoli, corn, cucumbers, eggplant, melons, peas, peppers, pumpkins, squash, strawberries, tomatoes

Root: beets, carrots, celeriac, garlic, onions, parsnips, potatoes, turnips

Note: This classification can be helpful as these crop groups have many similarities in nutrient and cultural needs (weed control, planting, and harvesting).

Crop	Seeding date	Days in the field	Family	Planting size	Nutrient needs	Harvested part
Beans	M, L	90	legume	large	low	fruit/flower
Beets	E, M	90	beet	small	medium	root
Broccoli	E, L	90	crucifer	medium	high	fruit/flower
Cabbage	E, M, L	90	crucifer	medium	high	leaf
Carrots	E, M	90	carrot	small	medium	root
Celery/celeriac	M	120	carrot	small	medium	leaf/root
Chard	E, M, L	60	beet	small	medium	leaf
Corn	M	90	grass	large	high	fruit/flower
Cucumbers	M	90	cucurbit	small	low	fruit/flower
Eggplant	M	120	nightshade	medium	medium	fruit/flower
Endive	E, M, L	60	lettuce	small	medium	leaf
Garlic	L	90	allium	small	high	root
Kale	E, M, L	90	crucifer	medium	high	leaf
Lettuce	E, M, L	60	lettuce	medium	medium	leaf
Melons	M	90	cucurbit	small	low	fruit/flower
Onions	E, L	90	allium	small	high	root
Parsley	E, M	120	carrot	small	medium	leaf
Parsnips	E	120	carrot	small	medium	root
Peas	E, L	90	legume	small	low	fruit/flower

Table 1. Cash-Crop and Cover-Crop Characteristics for Grouping.

Peppers	M	120	nightshade	medium	medium	fruit/flower
Potatoes	E	120	nightshade	medium	medium	root
Pumpkins	M	120	cucurbit	large	low	fruit/flower
Spinach	E, L	60	beet	small	medium	leaf
Squash	M	120	cucurbit	large	low	fruit/flower
Strawberries	E	2 years	rose	medium	medium	fruit/flower
Tomatoes	M	120	nightshade	medium	medium	fruit/flower
Turnips	M	90	crucifer	medium	high	root
Barley	E, L	90 or 120+	grass	variable	low	
Clover	E, M, L	90 or 120+	legume	variable	low	
Fescues	E, M, L	90 or 120+	grass	large (pasture)	medium	
Millet	M, L	60, 90, or 120+	grass	variable	high	
Oats	E, M, L	60, 90, or 120+	grass	variable	medium	
Rye (cereal)	E, M, L	90 or 120+	grass	variable	low	
Ryegrass	E, M, L	60, 90, or 120+	grass	variable	high	
Sorghum-Sudan grass	M, L	90 or 120+	grass	variable	medium	
Vetch	M, L	90 or 120+	legume	variable	low	

Table 2. Cover-Crop Management.

Species	Best time to sow	Winter-kill?	pH range	Tolerance	Days to first mow[1]	Mow[2]	How to kill	Sow with . . .	Comments
Annual ryegrass	eSp, lSm, F	zone 4	6.0–7.0	wet, shade	45	2–3 times @ 6"–10"	mow & till	clovers, inter-sow into corn	quick weed competition
Barley	Sp, F, W	zone 6	6.0–8.5	heat, drought	45	2–3 times	mow in flower		tough to till in
Oats	eSp, lSm	yes	4.5–6.5	good land is best	45 on		mow	clovers, peas, vetch, favas	great winter-kill mulch
Cereal rye	eSp–W	no	5.0–7.0	drought, shade	45	2–3 times @ 6"–10"	mow in flower	vetch	next winter establisher
Buckwheat	lSp–lSm	yes	5.0–7.0	good land is best	15 on		mow		fastest weed competitor; P catch-crop
Sorghum-Sudan grass	lSp–lSm	yes	6.0–7.0	heat, drought	45	once @ 15"–20"	mow in flower	clovers, peas, vetch	great biomass producer
Japanese millet	lSp–lSm	yes		wet, drought, heat	45	2–3 times @ 6"–10"	mow in flower	clovers, peas, vetch	quick weed competition
Fescue mixtures[3]	Sp, lSm–eF	no	4.5–7.5	drought	45	at 8"–10"	mow & till	forage, sod grasses for all-season hardiness	reseeds itself, great biomass builder, forage

Nitrogen-fixing legumes

Berseem clover	Sp, ISm–eF	yes	6.5–7.5	heat, drought	40	2–3 times	mow & till	intersow in spring crops frost seed	fast-establishing legume
Cowpeas	eSm	yes	5.5–6.5	heat, drought	30		mow frost free date	intersow after competition	quick weed
Crimson clover	eSm, ISm	no	5.5–7.0	shade	40	3–4 times flower	mow in into crops	spring intersow too short, 6"	don't mow
Field peas	eSp, ISm–F	yes	6.0–7.0	good land is best	30		mow like oats	grain nurse crop incorporation	easy spring
Hairy vetch	eSp, eF	no	5.5–7.5	good land is best	40	2–3 times	mow in flower	sow with cereal rye in fall	good mulch, N producer
Red clover (short-perennial)	eSp, ISm	no	6.2–7.0	shade	40	2–3 times	mow & till	best legume to intersow overall	best forage clover
Sweet clover (yellow)	Sp–Sm	no	6.5–7.5	drought	60	once	mow in flower	frost seed into small grain	long taproots penetrate well
White clover (perennial)	IW, Sp, eF	no	6.2–7.0	drought, shade, wet	60	many times	mow & till	small grains as nurse crop	great for roads, pathways
Alsike clover (perennial)	IW, eSp, ISm	no	6.5–7.0	wet	60	once	mow & till	frost seed into sm. grain	does well with heavy, wet soil
Fava beans	eSp, ISm	yes	5.5–7.3	good land is best	50		mow in flower	grain & viny legume covers	good soil penetrator

1. Days to first mow are rough estimates based on UCSAREP research and author's experience. Variables of seeding date, fertility, water, and species variation can greatly affect the maturation of a crop. These estimates can be used as minimum time periods to establish a beneficial cover crop.

2. When mowing covers, a flail mower does the best job of reducing biomass to a small, incorporatable size while leaving the residue where you want it. Rotary or sickle-bar mowers also work moderately well when two cutting passes are made at different heights and in different directions.

3. Northeast pasture/sod fescue mixtures generally include sheep, hard, fine-leaved, and red fescue species as well as ryegrass and bluegrass.

Include Cover Crops in the Groupings

Including cover crops at this grouping stage is vital to making them an integral part of the rotation. Most farmers concentrate on providing the best land for their cash crops first, plugging cover crops into the mix where and if they can. This usually means few get included in the rotation, and the ones that do are an afterthought, almost guaranteeing they will not be managed well during the rush of the season. If financially possible, setting aside even one quarter of your land in cover each season—or, at the very least, limiting your operation to single cropping with a cover crop inter-sown with or immediately following each cash crop—will improve your soil. Growers with less economic pressure and a larger land base may be able to sow half of their rotated land to cover crops each year—vastly improving their soils over a shorter period of time. Land that is sown to cover crops every other season tends to be better "cleaned" of insect pests, diseases, and weeds than is land cropped year after year and sown to covers only over the winter.

Using the cover-crop management guide in table 2 (pages 26–27), one can gauge where during the season there's time to sow, mow, or till in covers—planted before, after, alongside, or within the same units as your cash crops. The benefits of fitting in summer covers are very high: these hearty crops make big contributions to organic matter, nitrogen fixing, habitat for beneficial insects, and the general diversity of your farm.

Red clover.

Take care in mowing and tilling cover crops adjacent to cash crops, as the removal of habitat that the cover provides can force pest populations living there to move into cash crops. It helps to manage blocks of covers in strips, mowing/tilling only half the block at a time. This allows pest populations living in that block to migrate within the cover crop rather than to your adjacent cash crop. Another strategy is to till or mow when the stage of growth of adjacent cash crops makes them less susceptible.

Compiling Your Crop Groupings and Adding Them to the Map

Where to Start and What to Consider

The next step, using these different lists, is to group crops that work together in a multiyear rotation, both within your individual units and on your farm as a whole.

Just thinking about this step can deter growers with complex, diversified operations from attempting rotation planning. But it will help if, rather than making hard-and-fast groupings that move perfectly from unit to unit around the farm like pieces on a game board, you paint with wide strokes and plan on making compromises from the start. Use several extra copies of your farm map(s) for this process.

When finding places on your map for crops, it usually makes sense to set down larger-acreage crops first. Gradually plug all your crops into the units according to the priorities for your farm.

Including cover crops at this point in the planning can be helpful, as they can fill unused space in the units and begin to settle into your rotation where they can be most effective. Here are a few points to consider:

- Pair cash crops with cover crops from the fall before. All of the cover crops that winter-kill (oats, peas, millet, sorghum-Sudan grass, etc.) make the planting of early crops easy, as the dead residue incorporates quickly (see table 2 for covers that winter-kill).
- Each fall, you can maximize springtime flexibility for yourself by sowing as much available land as possible to winter-killed covers. Winter-hardy covers that keep growing in the spring can slow

the planting of the next crop during their decomposition. They can, however, preserve moisture and provide weed suppression by acting as mulch, and leave nutritive benefits after being tilled in.

- Generally, fields are workable in time to till in winter-hardy cover crops by early May, making most fields sowable by early June.

There are many ways to mix and match your crops with each other and to work covers into the rotation, so plan on trying several variations before you find ones that meet your objectives. Record identifying information like seeding dates ("lettuce, 8/30") and acreage needed ("450 bed ft., 0.12 ac.," etc.) for each crop, which will help ensure that all the crops in each unit will work together.

As you begin to see how crops fit on your map, consider the unique features of your farm. Concerns about irrigation access, spring wet spots, and fertility that were noted when the map was made should be considered in relation to the crops you have laid out. Production issues like access to the crops for the crew—or customers, in the case of U-pick crops—can also be considered at this point. Jean-Paul Courtens of Roxbury Farm in Kinderhook, New York had this thought: "One of the concerns with my rotation is keeping my greens crops close to the barn. We harvest them every day, and traveling out to the farthest field with the whole crew is not a good use of time if we want to get anything else done." If some of the physical features of your farm or production restrict rotating certain groups of crops over the whole farm, you may need to create two or more smaller rotations to accommodate the variations.

With the crops in place, it is possible to add tillage and other management notes as well as the ballpark dates the work will need to be done. For example, a unit that will be planted to sweet corn between May 30 and June 15 could be harrowed by April 30 to allow for a month of bare-fallow weed control. It could be recultivated May 15 and before each successive seeding to kill the flush of young weeds right before seeding the corn. Clover and ryegrass could be overseeded at the last cultivation of each planting, early- to late-July. The planning notes might look something like this:

Sample notes, unit B — 0.3 ac.
harrow oats/pea residue 4/30
bare fallow
recultivate 5/15
harrow/sow corn 5/30, 0.1 acre
harrow/sow corn 6/7, 0.1 acre
harrow/sow corn 6/15, 0.1 acre
intersow annual ryegrass
red clover 7/1–7/30
flail-mow cornstalks 10/15

Making notes with such projected dates for every unit, including those that are solely cover crops or pasture, will allow you to create a rough schedule of fieldwork for the farm. Transferring these dates to a calendar can help identify the seasonal peaks of labor and be the basis for adjustments in the rotation that distribute tasks away from those times, improving the quality of the farmer's life along with the soil.

Sample Groupings/Rotations on a Hypothetical Farm

Tables 3, 4, and 5 (beginning on page 36) show three sample rotations involving cash crops and some of the associated covers from the groupings described above. These examples are for a diversified vegetable farm with 12 acres of rotated land. Each example represents a different level of detail, structure, and field diversity. Each likewise represents an attempt to separate the groupings by at least four years. The chief variation is the size of the rotational units used to divide up the 12 acres. The first example is a 6-unit rotation, each unit covering 2 acres and containing several groups of crops. The second and third examples use 12 and 24 units to divide the same 12 acres, giving unit sizes of 1 acre and ½ acre, respectively, that contain two or just one crop group.

The groupings of crops used in these sample rotations are listed below. These groupings were formed starting with their field seeding/planting dates, then broken down further in this order of priority: length of time in the field, planting size, family, part of the plant that's harvested, and nutrient needs.

Early root/leaf
Precede with winter-killed cover
beets 3/30, 4/15
carrots 3/30, 4/15
chard 3/30, 4/10, 4/20, 4/30, 5/10
endive 3/30, 4/10, 4/20, 4/30, 5/10
kale 3/30, 4/20, 5/10
lettuce 4/10, 4/20, 4/30, 5/10
spinach 4/10, 4/20
peas 3/30
Follow with millet/peas

Midseason roots
Precede with winter-killed cover
carrots 5/1, 5/15, 6/1, 6/15, 6/30
beets 5/1, 5/15, 6/1, 6/15, 6/30, 7/15
celery/celeriac 6/5
turnips 6/30
peas 7/15
Follow with rye

All-season roots
Precede with winter-killed cover
potatoes 3/30
parsnips 3/30
onion (storage) 4/5
parsley 5/15
Follow with rye

Strawberries
Precede with winter-killed cover strawberries 4/15
Winter squash/pumpkins
Precede with winter hardy cover

winter squash/pumpkins 6/5
Follow with intersown red clover

Late brassicas

Precede with winter-hardy grain/legume cover
broccoli 8/1, 8/15, 8/25
cabbage 8/1, 8/15, 8/25
kale 8/15, 9/1
No cover to follow

Forage

Precede with winter-hardy grain/legume cover
fescue/bluegrass 8/15

Late leaf

Precede with winter-hardy cover
chard 8/1, 8/15, 8/30
endive 8/1, 8/15, 8/30
lettuce 8/1, 8/15, 8/30
spinach 8/15, 8/20/ 8/25, 8/30
No cover to follow

Sweet corn

Precede with winter-killed grain/legume covers
sweet corn 5/15, 5/22,
sweet corn 5/30, 6/5, 6/15
Follow with winter-killed cover if possible

Hot crops

No cover/winter-killed cover preceding
peppers 6/5
tomatoes 6/5
eggplant 6/5
melon 6/5
cucumbers 6/5
Follow with rye/vetch

Winter alliums
Precede with winter-hardy cover
garlic 10/5
sweet onions 8/15

Midseason greens
Precede with winter-killed cover
chard 6/1, 6/20, 7/10
endive 6/1, 6/10, 6/20, 6/30, 7/10, 7/20
lettuce 6/1, 6/10, 6/20, 6/30, 7/10, 7/20
kale 6/1, 6/20, 7/10
beans 6/1, 6/20, 7/10
Follow with rye/vetch
20% of land to cover-crop fallow
    10% to oats peas
    10% to ryegrass/clover

Each rotation example is a snapshot of a single growing season on our hypothetical 12-acre farm. To maximize efficiency, ballpark dates for tillage, cover-crop management, and compost spreading are synchronized for multiple units.

The rotation is animated for succeeding years by moving the crops in each unit one unit down (6-unit example) or to the right (12- and 24-unit examples). Each crop inherits the cover crop left behind the previous fall when it enters a new unit. Cropping in each unit is limited to a single cash crop, and each model has three acres devoted solely to cover crops every year, in addition to an acre each of multiseason strawberries and forage. The annual crops "jump over" the units containing perennials except for every fourth year when strawberries and forage crops are plowed under at the end of June. New plantings of forage and strawberries at this point are already establishing themselves in new units sown and transplanted the previous August and April, respectively.

As the berries and forage enter and leave the active rotation every fourth year, time for both crops to establish, produce, and be plowed under is built into their tenure in the soil. This is a simple and effective way to include limited perennial crops within any annual crop-rotation

---

Another Option

Another option for including perennial crops in rotation is to sow/plant a fraction of the total acreage needed each year. The size of the fraction should be equal to the number of years that crop needs to be in the ground (e.g., one-third of the total acreage for crops needing 3 years). If strawberries are to get 3 productive seasons plus one to establish the crowns (4 years in the ground), we would start out planting one-quarter of the plants. By the fourth year, after picking, the first planting would be plowed down (and the newest planting would be established and ready for production). This model keeps the whole rotation active but means that the perennial crop is never going to be completely uniform in age or production strength.

---

system, provided there is an equal area of cover-cropped land to rotate with these crops.

The perennial crops in these examples serve as long-term structure and fertility builders for the soil, like covers. The forage crop does this through the multiyear development of its root structure (and animal manure, if in pasture) and the strawberries contribute organic matter by way of straw mulch that's added each year to protect the crowns in winter. These perennials are diagrammed together for simplicity in this model. In the field there may be many reasons not to grow these two crops next to each other, or to break up the plantings into several interspersed within the rotation.

The 6-Unit Rotation

The 6-unit rotation in table 3 (pages 36–37) is the first example and assumes each unit is 2 acres in size, which allows the farmer to locate several different crop groups within each unit. A rotation with large units gives the farmer leeway to move crops around within a unit in any given year, in effect creating a rotation within a rotation. This can be helpful for growers with springtime wet areas, variations in fertility, or other conditions that only some of their crops can tolerate. In addition, a "loose" rotation like this can be advantageous for a high-production retail/wholesale operation

Table 3. Sample 6-Unit Rotation Plan (see page 35).

Unit 1 (units 2 acres each, 4 subareas per unit)

Season				
Unit 1				
Holdover	oat residue	oat residue	ryegrass/clover	ryegrass/clover
Spring	3/30–4/5 harrow all-season roots	5/1 harrow	5/1, 6/25 mow	5/1, 6/25 mow
Summer		6/1–7/10 midseason greens	7/1 plow, bare fallow	7/1 plow, bare fallow
Late summer	rye	rye/vetch	8/1 late brassicas	8/1–8/30 late leaf
Fall			no cover	no cover
Unit 2				
Holdover	rye	rye/vetch	no cover	no cover
Spring	5/1, 6/25 mow	5/1, 6/25 mow	6/1 harrow	4/15 harrow
Summer	7/1 plow, bare fallow	7/1 plow, bare fallow, compost	6/5 hot crops	5/1–7/15 midseason roots
Late summer	compost; 8/15 oats/peas	8/15–10/5 winter alliums	9/15 rye/vetch	rye
Fall		mulch w/straw		
Unit 3				
Holdover	oat/pea residue	mulched rows	rye/vetch	rye
Spring	earliest harrow; 3/30 early root/leaf	3/15 rake off mulch	4/15 mow rye	4/15 mow
Summer	7/30 harrow	7/15 harvest garlic, 7/20 harrow	5/5 plow	5/5 plow
Late summer	8/1 millet/peas	8/1 millet/peas	6/5 winter squash (pumpkins too)	6/5 winter squash (pumpkins too)
Fall			intersow red clover	intersow red clover

Unit 4

Holdover	millet/pea residue	millet/pea residue	red clover	red clover
Spring	5/5 harrow 5/15 corn	5/12 harrow 5/22 corn	5/2 harrow, bare fallow 5/30 harrow	5/2 harrow, bare fallow 6/15 harrow
Summer			5/30 corn	6/15 corn
Late summer	mow/plow corn, oats	mow/plow corn, oats	intersow red clover/ryegrass @ last cultivation	intersow red clover/ryegrass @ last cultivation
Fall	compost 8/15 forage		flail-mow corn after harvest & covers established	flail-mow corn after harvest & covers established

Unit 5–fallow & transition to semi-perennial

Holdover	oat residue	oat residue	ryegrass/clover	ryegrass/clover
Spring	frost-seed yellow sweet clover 6/15 mow	frost-seed yellow sweet clover 6/15 mow	5/1, 6/25 mow	5/1, 6/25 mow
Summer	7/15 plow, bare fallow, compost	7/15 plow, bare fallow, compost	7/15 plow, bare fallow	7/15 plow, bare fallow
Late summer	8/15 ryegrass/clover	8/15 ryegrass/clover	compost 8/15 oats	compost 8/15 oats
Fall				

Unit 6–fallow & semi-perennial crop–in rotation every fourth year

Holdover	rye/vetch	rye/vetch	oat residue	oat residue
Spring	mow, mow	mow, mow	4/15 strawberry	4/15 strawberry
Summer	7/1 plow, bare fallow	7/1 plow, bare fallow	7/1 mow leaf	7/1 mow leaf
Late summer	compost 8/15 forage	compost 8/15 forage		
Fall	mulch w/straw		mulch w/straw	mulch w/straw

Table 4. Sample 12-Unit Rotation Plan (see page 41).

Season	Unit 1	Unit 2	Unit 3	Unit 4
Holdover	ryegrass/clover	no cover	rye	red clover
Spring	5/1, 6/25 mow	harrow before sowing & for weed control	4/15 mow 5/5 plow	5/2 harrow, bare fallow 5/30 harrow
Summer	7/1 plow, bare fallow, compost	6/5 hot crops 5/1–7/15 midseason roots	6/5 winter squash (pumpkins too)	5/30 corn 6/15 corn
Late summer	8/1 late brassicas 8/1–8/30 late leaf	rye	intersow red clover	intersow red clover/ryegrass @ last cultivation
Fall	no cover			flail-mow corn after harvest, covers established

Season	Unit 5	Unit 6	Unit 7	Unit 8
Holdover	clover/ryegrass	oat/pea residue	millet/pea residue	ryegrass/clover
Spring	5/1, 6/25 mow	3/15 rake mulch 3/30 early root/leaf	frost-seed yellow sweet clover, 6/15 mow	5/1, 6/25 mow 7/15 plow
Summer	7/1 plow, bare fallow, compost	7/15 harvest garlic	7/15 plow	7/15 plow
Late summer	8/15 oats/peas 8/15–10/5 winter alliums	7/30 harrow	bare fallow	bare fallow
Fall	mulch garlic w/straw	8/1 millet/peas	8/15 ryegrass/clover	8/15 oats

Season	Unit 9	Unit 10	Unit 11*	Unit 12*
Holdover	oat residue	oat residue	rye/vetch	
Spring	4/30 harrow, bare fallow 5/15, 5/22 corn	harrow 3/30–4/5 all-season roots	5/1, 6/25 mow	4/15 strawberries
Summer		bare fallow rows before greens		
Late summer		6/1–7/10 midseason greens	7/1 plow, bare fallow	7/1 mow leaf
Fall	mow/plow corn	rye & rye/vetch compost; sow oats	compost 8/15 forage crop	mulch w/straw

*Units 11 and 12 in rotation every fourth year

Table 5. Sample 24-Unit Rotation Plan (see page 41).

Season	Unit 1	Unit 2	Unit 3	Unit 4	Unit 5	Unit 6
Holdover	ryegrass/clover	no cover	rye	red clover	ryegrass/clover	no cover
Spring	5/1 mow 6/25 mow	4/15 harrow	4/15 mow 5/5 plow	5/15 harrow 6/15 harrow	5/1 mow, 6/1 plow, compost 6/20 buckwheat	6/1 harrow
Summer	7/1 plow, bare fallow, compost	5/1–7/15 midseason roots	6/5 winter squash (pumpkins too)	6/15 corn	7/20 mow 7/25 harrow	6/5 hot crops
Late summer	8/1–8/30 late-leaf		intersow red clover	intersow red clover/ryegrass @ last cultivation	8/1 late brassicas	9/15 rye/vetch
Fall	no cover	rye		flail-mow corn after harvest, covers established	no cover	

Season	Unit 7	Unit 8	Unit 9	Unit 10	Unit 11	Unit 12
Holdover	rye/vetch	red clover	oats/peas	millet/fava residue	clover/ryegrass	rye/vetch
Spring	4/15 mow rye 5/5 plow	5/1, 6/25 mow 7/15 plow	frost-seed alsike clover 6/15 mow	5/1 harrow, bare fallow 5/30 harrow	5/1 mow 5/5 plow	5/15 mow 6/15 mow
Summer	6/5 winter squash (pumpkins too)	bare fallow	8/15 plow	5/30 corn	6/1–7/10 midseason greens	7/1 deep-plow, bare fallow
Late summer	intersow red clover	8/15 oats/peas	compost 8/15 millet/favas	intersow red clover/ryegrass @ last cultivation	rye/vetch	8/15–10/5 winter alliums
Fall				flail-mow corn after harvest, covers established		mulch w/straw

Table 5. Sample 24-Unit Rotation Plan (continued).

	Unit 13	Unit 14	Unit 15	Unit 16	Unit 17	Unit 18
Holdover	mulched rows	oat/pea residue	ryegrass/clover	millet/pea residue	oat/pea residue	rye
Spring	3/15 rake off mulch 4/15 mow rye 5/30 mow rye	6/15 frost-seed yellow sweet clover, mow 7/15 plow	5/1, 6/25 mow 7/15 plow	4/30 harrow, bare fallow 5/15 corn	earliest harrow compost 3/30–4/5 all-season roots	5/1, 6/15 mow 7/1 plow
Summer	7/15 harvest garlic 7/20 harrow, bare fallow	bare fallow	bare fallow			bare fallow
Late summer	compost 8/15 oats/peas	8/15 ryegrass/ clover	8/15 millet/pea	mow/plow corn after harvest oats/peas	rye	8/15 oats/peas
Fall						
	Unit 19	Unit 20	Unit 21*	Unit 22*	Unit 23*	Unit 24*
Holdover	oat/pea residue	millet/pea residue	rye + legume	rye + legume	winter-killed cover	winter-killed cover
Spring	earliest harrow 3/30 early root/ leaf	5/1 harrow, bare fallow, 5/22 corn	5/1, 6/25 mow	5/1, 6/25 mow	4/15 strawberries	4/15 strawberries
Summer			7/1 plow, bare fallow	7/1 plow, bare fallow	7/1 mow leaf	7/1 mow leaf
Late summer	8/1 millet/peas	intersow red clover/ryegrass @ last cultivation	compost 8/15 forage crop	compost 8/15 forage crop		
Fall		flail-mow corn after harvest, covers established			mulch w/straw	mulch w/straw

*Units 21, 22, 23, and 24 in rotation every fourth year.

that needs maximum flexibility throughout the season to plug in missed or new plantings where the soil is ready. By knowing the rough size of a crop grouping and the cover that needs to follow and precede it, this type of rotation can be quickly organized or reorganized during the season. If strictly kept, this rotation gives a minimum of four years between crop families. If disease or insect pests should loom, either problem could call for adjustments, making long-term maintenance of the rotation difficult. However, many northeast growers maintain similar rotations by recording which rows or beds each crop is planted in every year, allowing them to avoid planting families on the same land for several seasons. Some of these growers report they have maintained similar rotations for over twenty years without significant problems and have gained benefits in fertility, insect pest control, and weed suppression.

The 12-Unit Rotation

The 12-unit rotation shown in table 4 (page 38) provides more structure than the 6-unit version, containing two crop groups with similar timing in a single unit of 1 acre. Adding more structure than the 6-unit plan ensures that there are at least four years between crops of the same family landing in the same unit. Rotating into smaller units also means less thought is needed during the season to decide where each crop will be planted within each unit.

The 24-Unit Rotation

The 24-unit rotation outlined in table 5 (pages 39–40) contains ½-acre units, and only one grouping of crops is grown in each unit every season. This rotation can be labor-intensive to devise and takes a good knowledge of your land and a relatively fixed growing plan to establish. The advantage is that it maximizes succession variety and field diversity in cash and cover crops from one ½-acre unit to the next because of the patchwork or checkerboard of variation it allows. Because of this diversity and the built-in assurance that no cash crop will be grown in the same spot for at least four years, it is probably the most beneficial of the three rotation models with regard to soil health. The difficulty is that it assumes similar soil fertility along with equal access to fields and infrastructure (irrigation, harvest, etc.) during all seasons.

Once established, this rotation model is like the cruise control of farm planning. Detailed rotations create breathing room in the season to look around and see details of the farm with a new eye, relieved of worry about where that errant lettuce planting is going or how much cover crop seed to order each month.

By limiting the unit size to ½ acre, this model splits large crops into several units and spreads them out over the whole farm. This can serve the goal of insect pest and disease control by physically isolating these problems to just a part of the year's crop. The benefit can be expanded to smaller crops as well, by planting each succession within the season into a new unit, isolating its disease/pest problems to only one planting. One northeast grower separates her tomato successions to different areas on the farm, lessening the spread of disease problems this crop commonly has at the end of the season. The flip side of having one crop in multiple units is the extra attention it requires to monitor, manage, and harvest in different areas of the farm.

Experienced farmers who know their land and have relatively uniform soils can more easily undertake a more detailed rotation, while less experienced growers or those on more varied land may find a "loose" rotation best for them.

These three models also illustrate how a rotation plan might be gradually introduced into an operation. A farmer might begin with the 6-unit and move toward more detail and diversity in the 12- and 24-unit examples.

Whatever effort you commit to, it's important to remember that truly functional rotations are flexible in design, allowing for substitution of crops as needed by market, weather, and labor changes without derailing the overall rotation. Like a good road map, a rotation plan should include a number of alternative routes to the same destination. It's better to make compromises when designing—and prepare to make many more out in the field—than to throw the whole plan out the window or never make the effort in the first place. The process of thinking through your crops, mapping and measuring your acreage, and thinking ahead several years will make these deviations into better decisions. Like the goals of sustainability, improved soil, and greater profits, perfecting a rotation on your farm is something to work toward over the course of a lifetime.

five

The Economics of Rotations and Cover Cropping

The preceding chapters emphasize the soil and ecological improvements associated with a good rotation and cover crops. While these cultural arguments are hard to refute, a look at the financial benefits can really settle the decision. This chapter explores the capital outlays of establishing these systems, their production costs and savings, and the bottom line.

Successful organic growers, like other enduring farmers, are shrewd entrepreneurs. There is little benefit to all of this hard work if it doesn't feed your family and keep a roof over your head—and one on the barn, for that matter. A rotation system that includes cover crops can offer a real payoff for those willing to make the investment and wait a few seasons. At minimum you'll likely experience savings of time and a better cash flow for supplies and seed as you incorporate rotation thinking into your farming.

Organizing Your Work

Time is precious for the northeastern farmer—especially between May and October. Time management may be one of the most important skills for a grower to have. It may also be the least expensive to develop. Rotations are, apart from their biological uses, primarily organizational tools that can save time during the heat of the season. Having a picture of the workflow, from fall cover crop to spring tillage to cash crops to the next cover crop, allows the grower to think one step ahead. Add the seasonal specifics of seed/plant quantities, spacings and dates, amendment rates, tillage types, irrigation needs, and any other minutiae that might otherwise pull you out of the field—and you have something of real value. Knowing where things are going, how much is going there, what it needs to meet its yield, and what is going in afterwards creates peace of mind and makes it possible to devote more thought and energy to other perhaps more profitable or life-enhancing matters.

Reducing Inputs and Labor through Soil Improvement

Creating a rotation for your farm that includes cover crops improves the quality and yield of your cash crops largely by building the health of your soil. In addition to the marketing benefit of better crops, healthier soil reduces the yearly inputs (and labor) to support them. Although most organic farms strive for some degree of sustainability, they spend a considerable amount of cash on inputs. Costs of field amendments (compost, minerals, NPK fertilizers) and pest/disease controls can add more than a blip to the operating expenses of a farm. In fact, many growers—especially those with intensive cropping plans—commonly spend almost twice as much on compost and amendments as on seeds and plants for the field.

Over time, the influence of a rotation and cover cropping on the health of the soil allows farms to become much more self-sufficient. With good tilth, amendments are retained better in the soil, thus remaining more available to crops. Timely cover cropping helps to capture valuable soluble nutrients, especially nitrogen, and keep them from leaching into groundwater. Healthy soils also maintain the balance of microorganisms, keeping at bay those that cause disease and reducing the need for purchased disease controls.

Insect pests dwindle in numbers when their favorite crops move around the farm, avoiding overwintered populations emerging in the spring. And healthy soil grows plants that are more resistant to whatever damage pests can inflict.

All the savings on amendments and controls once a rotation settles in are matched by the labor hours saved in not applying them. Fewer trips to the fields with the spreader or spinner and less time running the crew with backpack sprayers adds up very quickly, especially when you consider that these tasks usually fall during crunch times of the season.

One grower cites his rotation as a big cost saver in control of the Colorado potato beetle. Rotation combined with good life-cycle information on the pest make for an effective strategy for its control. Knowing that the first generation of beetles must walk to their food source from their winter hiding places, he moves his potato crop at least 500 yards from the last planting, ensuring that fewer of the overwintered beetles will find it. By knowing where his potatoes were previously, he can make a good guess as to which edge of this season's crop the beetle will attack first. Closely monitoring this border allows him to control this first generation (and their larvae) by handpicking the bugs or localizing one or two sprayings of Bt before they get established. Savings in labor and the control itself are calculated in this way:

"Pre-rotation" thinking
Time to apply controls to 1 acre
  of potatoes using a backpack sprayer   =   4 hours
Labor costs @ 10/hr.*   = $40
Cost of Bt control per acre   =   <u>$35</u>
                  $75
Total cost for two 1-acre applications   =   **$150**

"Post-rotation" thinking
Time to handpick beetles on 0.1 acre   =   .5 hours
Time to apply controls on .1 acre
  with backpack sprayer   =   .5 hours
Labor costs @ $10/hr.*   = $10.00
Cost of Bt control for .1 acres   <u>$3.50</u>
                  $13.50

Total cost for two 0.1-acre
  pickings and applications   =   **$27.00**

*Hourly rate includes farmer costs of FICA, Medicare, worker's compensation insurance.

Costs of Establishing a Rotation with Cover Crops

While the benefits of establishing a rotation that includes cover crops can be significant, the costs involved with creating and maintaining it season-to-season should be considered. There are hard costs in addition to the conceptual effort of corralling your operation into an organized system. These include taking land out of cash-crop production, possible extra time (at least initially, or if adding cover crops), cover-crop seed and management costs, and capital expenses for management equipment. The rest of this chapter explores some of the questions and costs growers might come up against when starting a rotation with cover crops.

Land Costs

Using cover crops systematically takes land out of cash-crop production. Depending on the level of commitment you decide to make to the rotation and the intensity of your cash cropping, the changes you make can be large or small. A grower who double-crops almost all available acreage, with barely enough time to sow a winter cover on some of the land, is going to have a tougher time with this decision than one who has some land already fallow or is just single-cropping. Questions to ask to help determine the value of cover cropping for your farm might include these:

- What is the income I am bringing in per acre, per year, or for each harvest (if more than one per year)?
- How much am I spending on inputs to maintain this level of production per year? (Include fertilizer, pest and disease controls, and the labor to apply them.)
- What would be a tolerable level of reduced production if offset by a corresponding reduction in input costs over many years?
- Or, in contrast, is there affordable land available to expand the rotation onto?

These financial questions can be contrasted with some other queries about the long-term health of the farm:

- Looking at soil analysis for the farm, can I see a change in the numbers that might relate to the amount of inputs every year? Is this change positive, negative, or mixed? (An example would be the farmer who adds chicken manure every season to keep production high and doesn't use a cover crop to "catch" excess fertility after the cash crop. This grower also sees a gradual rise

in phosphorus, salt, and pH levels in the soil, maybe to the point that they are out of balance with sustainable soil health).
- Does the health of my crops seem to be improving, declining, or maintaining itself each season using my present systems?
- Do weed and pest pressures on the farm seem to be improving, getting worse, or staying the same over many seasons?

While several of the questions above have answers that are not numbers, they can tip the scales one way or another in assessing the long-term value of taking land out of production to start a rotation on your farm. The right rotation plan will make it possible to balance good horticulture with production needs, keeping the farm in the black. The more thinking and detail a farmer can bring to creating a rotation, the greater the value it can deliver to the operation.

Cover-Crop Seed Costs

Table 6 lists several cover crops popular on northeast farms along with their seeding rates and cost per acre to sow.

Also included in the table are the types of inoculants to be added to each legume crop when sown. Inoculating the seed before sowing increases its ability to fix nitrogen in the soil. The cost is usually minimal, and seed suppliers can deliver them with the order. (Inoculants need only be added if inoculated legumes of the same variety have not been grown in the same field for more than three years. Be sure to check with your seed supplier and certifying agency for organically approved, non-GMO inoculants.)

Management

Managing covers during the season can take time away from managing cash crops. Understanding the time investment needed to sow, mow, and till in a cover is important when planning where that cover will fit within a rotation. Table 2, "Cover-Crop Management," gives a general idea of which covers to sow, how long until they might need mowing, and how many times to mow them before tilling. See table 7 for an example of the costs of managing two acres of covers for one season. Given an understanding of how often you will need to be in the field managing the crop, you can make general time estimates needed for each plot or sowing of

Table 6. Cover-Crop Seed Costs.						
Cover crop	Lbs. broadcast-seeding rate per acre[1]	Lbs. when sown with 2nd cover	Inoculant type	$ cost per pound	$ cost per acre alone[2]	$ cost per acre with 2nd cover
Annual ryegrass	25	12		$0.65	$16.25	$7.80
Barley	100	35		$0.35	$35.00	$12.25
Oats	125	50		$0.13	$16.25	$6.50
Cereal rye	125	50		$0.15	$18.75	$7.50
Buckwheat	75	25		$0.30	$22.50	$7.50
Sorghum-Sudan grass	45	n/a		$0.70	$31.50	
Japanese millet	55	30		$0.80	$44.00	$24.00
Fescue mixtures[3]	variable	variable				
Nitrogen-fixing legumes						
Berseem clover	20	8	clover	$2.40	$48.00	$19.20
Cowpeas	100	40	cowpea	$1.30	$130.00	$52.00
Crimson clover	27	10	clover	$1.70	$45.90	$17.00
Field peas	95	40	pea	$0.45	$42.75	$18.00
Hairy vetch	35	25	vetch	$2.20	$77.00	$55.00
Red clover (short perennial)	15	10	clover	$1.30	$19.50	$13.00
Sweet clover (yellow)	15	n/a	alfalfa	$1.08	$16.20	
White clover (perennial)	10	6	clover	$2.30	$23.00	$13.80
Alsike clover (perennial)	10	5	clover	$1.20	$12.00	$6.00
Fava beans	170	80	pea/vetch	$1.05	$178.50	$84.00

1. If the seed is sown with a grain drill, the quantity can be reduced by 25%–50% with the same germination result.
2. Price sources: Corland Seed Co., Peaceful Valley Farm Supply, and Ernst Seed.
3. Fescue mixtures, their seeding rates, and price per pound vary greatly depending on their uses.

Table 7. Costs of Managing Two Acres of Cover Crops for One Season.				
Crop	Task	Labor cost*	Fuel cost	Task total
1 acre rye/vetch	3 mowings	3 hours@$10/hr.	$30	$60
	till/prep	1 hour@$10/hr.	$10	$20
1 acre rye/vetch	sowing	1 hour@$10/hr.	$10	$20
1 acre ryegrass/ clover	3 mowings	3 hours@$10/hr.	$30	$60
*Labor hourly rate includes farmer costs of FICA, Medicare, and worker's compensation insurance.				

covers during the season. Add a rate for machine work that includes fuel and/or labor costs, and a clear picture emerges of what the annual costs of growing covers for your farm might be. You can judge whether this cost figure, against the long-term soil benefits cover crops offer, represents one you are willing to incur.

It's important to manage cover crops so they don't have a detrimental effect on your cash crop. For example, allelopathic chemicals left in the soil by some covers may inhibit germination of the cash crop. Also, the mowing or tillage of a cover may destroy a pest habitat, causing those pests to attack your cash crop.

Allelopathic chemicals are natural germination inhibitors that are produced by some plants to control competition from other species around them. The four cover crops that are known to produce them are hairy vetch, cereal rye, sorghum-Sudan grass, and oats. The chemicals can remain active in the soil for up to four weeks after the crop is killed/tilled-in. Understanding the extra time needed and including it in your rotation plan will allow you to work around allelopathic effects on your cash crop. It is also possible to use this chemical to your advantage by sowing these covers before a tough annual weed emerges, allowing them to suppress its germination during the period they are growing. (Note that transplanted or large-seeded vegetables are immune to allelopathy, which inhibits only small-seeded crop plants and weeds.)

Poorly timed work with a cover crop can also indirectly damage cash crops nearby. This usually happens when the cover crop is supporting a population that would otherwise be a pest for cash crops. Many of the grains and clovers used as covers harbor pest species at some point in the season. (See table 7.) When the foliage of the covers is removed or destroyed, these bugs are forced into the surrounding area to find other

Table 8. Covers and the Beneficial Insects and Pests They Can Harbor.		
Cover crop	Beneficials	Pests
Oats	ladybugs	leafhoppers, thrips, armyworms
Cereal rye	ladybugs, hover flies	armyworms
Buckwheat	parasite wasps, ladybugs, flowerbugs, various flies	tarnished plant bugs
Cowpeas	parasite wasps, ladybugs	stinkbugs
Crimson clover	ladybugs, big-eyed bugs	thrips, tarnished plant bugs
Hairy vetch	ladybugs, flowerbugs	cutworms, tarnished plant bugs

food and habitat. They'll readily settle into a neighboring cash crop that fits their needs, and perhaps do quite a bit of damage. Of course the age of the crop is an important factor, as older crops are usually more able to resist insect damage. It's often possible to avoid crop damage by managing blocks of covers in succession, allowing pests to move from one mowed/tilled strip to another one that's adjacent and still leafed out.

Cover crops also harbor beneficial insects and can be used to provide shelter and feeding areas for these "good bugs." Providing this shelter keeps them close to your cash crop where they can control outbreaks of insect pest species.

Capital Costs: Equipment for Growing Cover Crops

Most cover crops are entirely different from cash crops in their physical structure and their cultural needs. Understanding the process of growing each cover from sowing to tilling-in is vital to knowing how effective each will be with the rest of your rotation.

When sowing covers, the ability to maximize germination and minimize time spent is important. Most covers can be seeded in two ways: by broadcasting and with a grain drill.

Broadcasting usually presents less of an investment than purchasing a grain drill, as broadcasting seed can be done with either a hand-cranked spreader strapped to the farmer's chest or by a tractor-mounted PTO spreader. A broadcasting spreader is vital for intersowing cover crops into cash crops because it distributes seed evenly and efficiently over the top of the already-established crop.

The downside of broadcasting cover crops is that it requires a follow-up pass over the field with a roller or shallow harrow for good seed-to-soil contact. In addition, if you are sowing two covers together, unless the seed is similar in size, you usually will have to sow them independently, requiring two passes.

The benefit of a grain drill, especially one with a grass box for small seed, is that with one pass you can sow two types of seed, setting it to whatever depth you like while ensuring good soil contact. In addition, some types of drills also have fertilizer boxes that will allow you to amend and sow with one pass through the field. The downside is that they are more expensive than either a hand-cranked or tractor-mounted broadcaster. This difference can often be offset over the course of many seasons by having to make fewer passes than when broadcasting, as well as reduced seeding rates to achieve the same stand density.

In the best of all worlds, a grower would have a grain drill—to maximize efficiency when seeding large plots—as well as a broadcaster to intersow covers into cash crops already established.

Once sown, most cover crops, with the exception of those that winter-kill, will need to be mowed at least once. Mowing at the correct time promotes regrowth of root and leaf systems, enhancing their soil-enriching potential. Well-timed mowing is also the best way to deal with covers before tilling them in. Cutting their dense foliage into small pieces speeds the decomposition process after tillage. Equipment suited to this task can

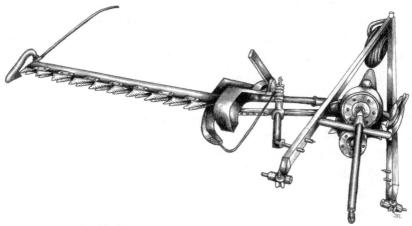

Tractor-mounted sickle-bar mower.

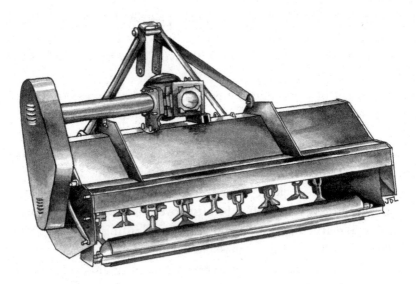

Tractor-mounted flail mower.

make working with cover crops a much more manageable experience. Many farmers, even those with a lot of experience, overlook this detail and spend many seasons waiting impatiently for cover crops to decompose before they can plant each spring.

Most farms have a mower of some type, the most common probably being the rotary type. There are also disc mowers out there as well as sickle bars. All three cut in pretty similar ways: horizontally taking down the foliage or stalks near the base of the plant and either laying them down where they stand or pitching them to the side. These mowers work well in creating sheet composting mulches with the dead covers, but unless multiple passes are made at different heights, they leave behind long stems that break down slowly while the next planting waits.

By contrast, the flail mower cuts into a cover crop vertically, its multiple blades spinning into a 12- to 18-inch-tall swath of growth from the base up the stalk. This vertical cutting action chops the whole height of the crop into pieces as small as 1 inch. Flail mowers discharge the chopped fiber right where it was growing, creating an even, thick mat that also works well as a sheet compost or mulch, but decomposes quickly when incorporated.

Jim Crawford of New Morning Farm in Hustontown, Pennsylvania,

Table 9. Equipment for Cover Cropping.				
Tool	Size	HP rating	Price new	Price used
Sowing				
Hand-cranked broadcaster	2 gal.		$35	
Tractor-mounted broadcaster	30 gal.	25	$550	$150
Grain drill	10 ft.		$3,300	$1,500
Mowing				
Bush-hog	4 ft.	25	$1,200	$600
Sickle bar	5 ft.	25	$2,100	$900
Flail	5 ft.	25	$2,400	$1,500
Incorporation				
Moldboard plow	1 bottom	25	$850	$400
Rotovator	40 in.	25	$1,800	$1,200
Disc harrow	48 in.	25	$2,200	$600
Chisel plow	3 shank		$1,400	$950
Subsoiler	1 shank		$560	$400
Reciprocating spader	47 in.	40	$3,700	n/a
Sweep-type field cultivator	48 in.	25	$2,600	n/a

says he "just about gave up on sowing rye in the fall after missing several planting dates in the spring waiting for it to decompose after mowing with my rotary bush-hog. I really underestimated the job a flail mower could do in taking down heavy cover crop growth."

Table 9 lists cover-cropping equipment and compares current new and used prices of several mowers and other pieces of machinery.

After the cover crop has been sown, has grown, and has done its job, incorporating it into the soil quickly and completely is the next task. Almost any tillage tool will do a decent job incorporating most of the cover crops grown. The key to maximum ease and efficiency is using the right tool and the right timing to leave the soil ready for the next crop. Plows, harrows, chisel plows, rotovators, spaders, and modified field cultivators all move through the field differently, turning in crop residues shallowly or deeply, incorporating large amounts of biomass with or without clogging up. Fully matured cover crops that make a lot of biomass are much easier to incorporate deeply if they have been allowed to wilt down first. Covers that are to be incorporated shallowly generally go under better when they are green and tender. Knowing what cover crops your tillage tool works with will guide decisions about any other tool to help with some of the covers you will be sowing.

six

Two On-Farm Examples

Eileen Droescher, Ol' Turtle Farm, Easthampton, Massachusetts

Land in Rotation: 12 Acres

For the membership of her CSA, Eileen grows an extremely diversified mix of vegetable and fruit crops on 14 acres in the Connecticut River valley. Of those, 12 are in rotation with the other two planted to cane fruit. She has devised a rotational plan that allows her to maximize production levels on her farm (supplying food for 200-plus families through the CSA) while still meeting her goal of devoting half of her rotated acreage to cover crops during each season.

The fields at Ol' Turtle Farm are organized in a permanent bed system and tilled with a spader that is the same width as the beds. This system allows Eileen to know exactly where a crop was sown in any given year and till in covers and cash crops one bed at a time—ensuring a seamless succession between the two. Her rotation is based on a few large units: 30 beds (each one 300 ft. long) make up a single unit. For diversity in the field, Eileen has designed her plantings so that each unit of 30 cash-crop beds is separated from the next by 30 beds of fallow covers. This patchwork pattern allows for beneficial insect habitat next to cash crops and helps maintain a protective distance against disease and pests between successive plantings of the same cash crop. Two examples Eileen cites are squash and tomato that are susceptible to mildew/blight problems as the season progresses.

Using large units subdivided by permanent beds allows Eileen to create a rotation *within* each one in any given season, varying the locations of crops easily among the 30 beds. This flexibility also maximizes her ability to change details like the size of a planting or its location for ease of harvest from year to year. Good record keeping and permanent beds allow her to avoid planting families in the same location for at least four years.

The basic rotation of covers and cash crops at Ol' Turtle Farm looks something like this:

Year 1: oats/peas
Year 2: cash crops
Year 3: sorghum-Sudan grass/red clover or millet/red clover
Year 4: cash crops
Year 5: rye or rye/vetch
Year 6: cash crops

David Colson, New Leaf Farm, Durham, Maine

Land in Rotation: 9 Acres

Dave and his family grow mixed vegetables for wholesale restaurant accounts as well as retail markets, mostly in the Portland area. Three acres out of nine are planted to cash crop each year, with the other six in cover crops. Increasing yields of high-quality greens and strong relationships with local chefs have provided them the ability to maintain revenue while increasing the acreage under cover each year. Intensive cover cropping allows them to aggressively build soil quality while at the same time minimizing weeds that can quickly overwhelm the family crew with hand weeding and increased harvest hours.

Many seasons of trial and error that included attempts at grain production and larger-scale cash cropping have brought New Leaf to its present rotation system. Matching intensive large-revenue cash plantings with two-season fertility-enhancing covers minimizes mechanical cultivation needs for weed control and labor needs for harvesting.

David's nine acres of fields are divided into four 2¼-acre sections. Two of these sections are sown to cover crop for two years at a time, one in red clover and the other in a succession of annual covers like rye/vetch and oat/pea. Another 2¼-acre section is sown to cash crops each year. The remaining 2¼-acre section is subdivided into three ¾-acre subsections, two sown to annual cover crop and the third in cash crop. The latter subsection remains in a mixture of cash and cover crop for two seasons and is then rotated into two seasons of cover. Dividing larger rotation

2.25 acres buckwheat → bare fallow →rye/vetch → bare fallow → oat/pea	2.25 acres cash crop

.75 acres annual cover crop	.75 acres cash crop	.75 acres annual cover crop	2.25 acres oats/red clover → red clover

Colson Rotation.

units into subsections allows David to create mini rotations within the larger system, giving him flexibility with his cash-crop plantings each sowing. The larger 2¼-acre sections of cash and cover rotate with each other either annually or biennially, depending on what covers are sown.

Leaving fallow sections in cover crops for two seasons maximizes the disease and pest suppression of the cover crops, allowing David to arrange plantings of cash crops around production criteria as opposed to strict crop-family succession. Having the clover develop for two seasons greatly boosts its nitrogen contribution to the following crop and also allows the root structure to penetrate to a greater depth, opening soils for less aggressive vegetable root systems. Rotating the annual covers for two full seasons, with bare fallows at varying points, quickly cleans up problem perennial and broadleaf weeds by disrupting their life cycle at every possible point.

Rotations, Cover Crops, and Green Manures

The following farmers' accounts are excerpted from chapter 2 of *The Real Dirt: Farmers Tell about Organic and Low-Input Practices in the Northeast*, Second Edition (1998). Edited by Miranda Smith and Elizabeth Henderson, *The Real Dirt* is another NOFA publication created with funding and collaboration from SARE.

Rotations for Agronomic Crops

Farmers growing grains and beans get double duty out of their rotations. Diverse rotations diminish weed and pest problems. Farmers also can harvest the grain of beans, leaving residues to add organic matter to the soil.

John Meyer works with a flexible rotational system that is based on the following scheme:

5 to 6 years alfalfa
2 years corn, of which about one-fourth is underseeded to clover and one-fourth is underseeded to rye;
1 year soybeans;
1 year wheat or rye, underseeded with medium red clover
1 more year of medium red clover
1 year wheat or rye
1 year soybeans

John describes the way he handles the grains underseeded with clover: "I frost-seed clover with the spinner spreader in early spring in wheat or rye planted the previous fall. After I harvest the grain, most of the clover stays on through fall of the following year. I green-manure most of it and harvest hay from about a quarter of the area for my cows. At second cutting, I combine the clover hay for seed. That fall I plow and plant wheat or rye—whichever I didn't have before—or jump back into corn again depending on the amount of manure or compost available, the fertility, and the weed population. I won't grow wheat if there are a lot of perennial weeds. I grow more soybeans and wheat than corn, so sometimes I plant soybeans after wheat."

Rotating Vegetables with Vegetables

Vegetable growers generally devote a great deal of attention to the ways in which they rotate their vegetable crops.

Eliot Coleman has been working to refine a vegetable rotational plan for many years. He describes this plan and the rationale behind it in *The New Organic Grower*. In brief, Eliot's plan goes like this:

Potatoes follow sweet corn.
Sweet corn follows the cabbage family.
Cabbage family crops, undersown with leguminous green manures, follow peas.
Peas follow tomatoes.
Tomatoes, undersown with a non-winter-hardy green-manure crop, follow beans.
Beans follow root crops.
Root crops follow squash and/or potatoes.
Squash is grown after potatoes.

Paul Harlow has developed a very successful rotation for his vegetable crops. Normally he grows alfalfa for three years and follows it with six years of vegetables before going back to alfalfa. After alfalfa, Paul grows winter squash or pumpkins. The next year, he plants brassicas, harrows them in fall to break down the stalks, and plants tomatoes and peppers the following year. Tomatoes never follow tomatoes for five or six years to avoid disease.

Paul usually follows early lettuce transplants with buckwheat, letting it grow almost to seed and then plowing it and planting oats in the middle of August. He harrows the winterkilled oats in spring and plants onions. After that, he plants carrots. He doesn't use any manure or compost with his carrots but does use 500 to 600 pounds per acre of Sul-Po-Mag for potassium.

David Colson has developed a rotational plan that he says "works well for both disease and weed control." It runs like this:

Year 1: Crucifer family and lettuce
Year 2: Cucurbits, alliums, peas and beans, potatoes, spinach
Year 3: Wheat undersown with red clover
Year 4: Clover green manure

Rotating with Livestock

Julie Rawson uses pigs, turkeys, chickens, geese, and ducks in rotations with her fruit and vegetable crops. As an example of this system, Julie raised pigs on a field that had been in grass hay for many years. The pigs were on the field for a year before she used the area for vegetables. She says, "One season of pigs wasn't enough to build the fertility on this land. Next year, I'll divide it into three sections for chickens, putting crops and buckwheat in before and after the birds."

Tillage Choices

In preparing a field for planting either crops or cover crops, many organic farmers avoid using a moldboard plow, which inverts the top layer of soil. The moldboard is reserved for breaking a sod or turning particularly heavy or gravelly soils. Instead, many farmers prefer the chisel plow, which cuts deep furrows in the soil without flipping it over. The chisel plow preserves the existing soil structure, mixing in some of the organic matter from the surface while opening air passages through the soil. These passages facilitate water penetration and the decomposition of the organic matter.

David Stern says that switching from the moldboard to the chisel plow was "the single most radical change I made." David built his own chisel plow in 1985 because those available commercially required high horse-power to pull (15 hp per tine, generally with eight or more tines). The four-tined plow he put together can be pulled by a 65 hp tractor. He saw the effects right away: the wet soils dried faster. Combined with rotations, compost, and cover crops, the chisel plow helped make the clay soils less sticky and the sandy soils held water better.

Dairy farmer Earl Spencer uses his chisel plow to mix manure and corn stubble into the soil before preparing a seedbed with a disk harrow. Christoph Meier of Hawthorne Valley Farm says his soils are so varied, from loamy to gravelly, that he must treat each differently. He uses a moldboard plow on gravel soils and a chisel plow on light soils, before disking. A few farmers even use the chisel plow to break a sod, though they admit it requires many subsequent trips over the field with a disk before a crop can be planted.

Jack Lazor chisel-plows in fall to aerate and ridge the soil. The ridging leaves the soil ready for spring planting and reduces erosion. The chisel shanks are three times the heft of his cultivator and penetrate much deeper. His 12-inch shanks provide enough clearance to handle all summer growth. He uses twisted shovels to throw the dirt and bury the straw and stubble trash. "The stubble acts as an air and water wick into the soil," he says.

Tony Potenza, who grows grains, beans, and vegetables, talks about the need to adjust practices to field circumstances. He says, "If the soybean harvest is early, I go into the chopped stover the combine left behind with a Glencoe shank cultivator with wide sweeps and a leveling bar on the back. I drill rye, or wheat, and frost-seed my own medium red clover with a Cyclone seeder. The ground is rough. Left that way, it holds more snow and gives more protection for the new sprouts.

"If the soybeans are late, then I chisel plow in fall and follow in spring with corn or with soybeans again and go back to winter wheat with clover. Depending on how well rye survives the winter, I harvest the grain or use it as a green manure."

Common Cover Crops and Green Manures

The now-defunct New Alchemy Institute studied uses of cover crops and green manures on farms in this region for several years. Their Research Report #10 ("Cover Cropping and Green Manuring on Small Farms in New England and New York: An Informal Survey") is an excellent resource for farmers new to these practices as well as those interested in innovative crops and techniques. "Managing Cover Crops Profitably" (Sustainable Agriculture Publishing, 1992) is a useful guide to cover crops.

Farmers use a wide variety of crops for these practices. Common ones include winter rye, clovers, buckwheat, oats, perennial grasses, alfalfa, hairy vetch, winter wheat, field peas, Sudan grass, annual ryegrass, forage brassicas, fava beans, soybeans, millet, and even weeds.

Fertility from Cover Crops and Green Manures

Peter Young, who produces cream-topped pasteurized milk and sauer-kraut from homegrown cabbage, says, "Organic farmers rely more on legumes for their nitrogen than do other farmers. Nancy and I use clovers and alfalfa to build our soil N."

"Rotting legume sod gives the nitrogen my corn needs," says dairy farmer Jack Lazor. "I use a corn-oats-alfalfa rotation just as many do in the Midwest."

Tony Potenza says, "Most of my manure goes to the vegetables. For grains, I rely on green manures for fertility."

Earl Spencer rotates alfalfa/brome grass hay with corn. He plows the sod in fall and in May spreads manure, works it in with a chisel plow or tandem disks, and plants the corn. After final cultivation, he undersows the corn with annual ryegrass, which he broadcasts with a rear-mounted spinner spreader. He tried undersowing vetch, but rye is cheaper and his main concern is erosion control.

Fertility improves when cover crops and green manures are turned into the soil. There is no argument about this effect even though the causes of the action are not always clear. Some researchers state that legumes don't contribute significantly to long-term nitrogen stores, although they do increase the nitrogen uptake of grasses grown with or after them.

Most agronomists agree that, in addition to the direct contribution of nutrients made by the decomposing crop, the decomposition process itself adds to fertility. As Eliot Coleman explains, "Decaying organic matter can make available otherwise insoluble plant nutrients in the soil through the actions of decomposition products such as carbon dioxide and acetic-, butyric-, lactic-, and other organic acids. . . . Increasing the carbon dioxide content of the soil air as a result of the decomposition of plant residues increases the carbonic acid activity, thus speeding up the process of bringing soil minerals into solution. . . . Soil microorganisms are also stimulated by the readily available carbon contained in the fresh plant material, and their activity results in speeding up the production of ammonium and nitrate."[1]

The choice of green manure directly influences nitrogen supplies, at least in the short term. Biennial and perennial legumes generally peak in nitrogen contribution in their second or third year.[2]

Some crops contribute nutrients other than nitrogen. Buckwheat, for example, is known as a good phosphorus collector, while timothy's a good potassium scavenger.[3] Mustard returns high calcium levels, while other brassicas collect sulphur.

Interested readers can find excellent information about the nutrient contributions of various crops in "Plowdown—A Strategy for the Eighties,"[4] and "Green Manuring—Principles and Practice."[5]

Some farmers report they do not see a dramatic increase in organic-matter content as a consequence of using green manures and winter covers. They expected to, but haven't, they note. Perhaps they will be cheered to learn that academic research supports their observations.

Effects on Organic Matter

Cover crops and green manures do not increase the percentage of soil organic matter unless they are left in place for longer than a year.[6] Instead they replace most of what was lost when the previous crop was harvested. If farmers wish to increase organic-matter content with these practices rather than by adding composts or other organic materials, they have no choice but to leave the crops in place for more than a year without harvesting or include a number of green manures in the rotation.

According to Eliot Coleman, however, "The cycling of organic matter involved in the decomposition of cover crops may lead to improved soil structure even without an actual increase in organic-matter levels."

By combining a sod plow-down with strip composting, David Stern and Elizabeth Henderson saw a dramatic increase in organic matter and cation-exchange capacity (CEC) on a field they were bringing into production.[7] They had never grown any vegetable except corn in this field, says David. "We let it sit under a mowed grass crop for two years and then fall-plowed it twice. Early the following summer we replowed, spread lake weeds at 40 to 60 tons wet-weight per acre, and planted rye-vetch. By next spring, the organic-matter content had risen half a percent and the CEC went from 6.5 to 11. To maintain these levels, we will continue to rotate green manures and cover crops with vegetables."

Benefits of Combined Plantings

Farmers often combine plants for cover crops or green manures to offset a disadvantage of one of the crop species they are using. According to New Alchemy researchers, "Rye and vetch planted together can solve some of each other's problems. Like other grasses, rye has a fibrous root system, which binds the soil and helps prevent heaving of the more tap-rooted vetch. Rye stems begin to elongate early in spring, providing support for vetch vines. This allows good light penetration and air circulation for the legume and thus promotes vigorous growth. By mid-May, the vetch shades the ground heavily, which, in combination with competition for moisture and the allelopathic effect of the rye, contributes to weed suppression. Finally, when the crops are turned under, the succulent vetch assists with breaking down the rye. C and N are well-balanced, so the residues neither rob the soil of N nor release too much at once, but provide N slowly and steadily through the season."

Like many other vegetable growers, Robin Ostfeld and Lou Johns harvest about 50 percent of their crop in time to plant rye and vetch on those fields.

Brian Caldwell and Twinkle Griggs, who raise vegetables, raspberries, chestnuts, apples, hay, forage, and lambs (from a fifteen-ewe operation), also use a rye-vetch winter cover on their vegetable acres.

Establishing Green Manures and Cover Crops

Farmers new to cover cropping and green manuring sometimes report difficulty in establishing these crops. Practiced farmers offer the following tips:

For many legumes: soil pH of 6.5 to 6.8, just to the acid side of neutral, and both molybdenum and cobalt levels must be adequate. An inoculant also is important, particularly on soils where the legume cover has not been grown within the last few years. Nurse crops, such as oats, often are used to protect the seedling legumes.

With any new planting, adequate moisture is key. Timing the planting just before a rain is ideal, but irrigation is sometimes necessary for establishing summer-sown green manures.

Drilling is more reliable than broadcasting for good germination. When broadcasting, lightly cover the seed, either by going over the area with a shallow tilling harrow, or by raking.

Frost-seeding hardy plants such as clovers and grasses works well. The alternate freezing and thawing in early spring acts to work the seed into niches in the soil surface where it will germinate as soon as the soil warms.

John Meyer's method of preparing a field for alfalfa provides an example. Although heightening the risk of erosion, he plows in the fall, disks in spring, and uses a culti-mulcher or field cultivator after that. "I plant alfalfa with a Brillion roller-seeder," he says. "There's a row of cultipacker wheels, seeds drop, and then another row of wheels push seed in a little. I seed 12 lbs. per acre, sometimes with oats as a nurse crop. I drill the oats first and then go over them with the Brillion seeder. I have a field now where I spread alfalfa into wheat early in spring with snow on the ground."

David Holm broadcasts mammoth red clover into rye either in March or early April, depending upon the weather. He mows the field just before the rye goes to seed and says that this system stimulates the clover. "It comes back very thick and lush. It's a better legume crop than the vetch I broadcast over rye that's gone to seed."

David Stern and Elizabeth Henderson also frost-seed mammoth red clover into rye. "In late June, we flail chop the rye and use it to mulch squash. The clover takes over. The next year, we rotovate in May to incorporate the clover, then chisel plow, spread compost, and plant sweet corn," they say.

Undersowing and Intercropping

Farmers using cover crops after late crops choose one of two options. They harvest in time to plant the cover (winter rye can be planted as late as October 1 in northern areas, October 15 in Massachusetts and southern New York, and November 1 in New Jersey and Rhode Island) or they undersow a winter cover crop into their late-season vegetable crops.

Undersowing is commonly done by broadcasting the cover crop a few weeks before harvesting the cash crop, in order to allow the cover crop to become established enough to withstand some trampling. Because undersowing is done so late in the season, competition problems are avoided. At Rose Valley Farm, Elizabeth Henderson and David Stern undersow earlier, cultivating a crop twice before seeding the cover crop. They sow clover under corn, and clover, oats, or rye under fall brassicas, with the choice depending on their plans for the next season. Intercropped green manures, both annual and perennial, also are used in orchards where they provide habitat for beneficial insects and improve the soil.

Eliot Coleman has used a system of intercropping green manures with his late-produce crops for many years. Because he prefers nitrogen-fixing legumes to grasses for most circumstances, he has experimented with various crops, timings, and sowing methods to develop workable systems. Eliot plants a cover crop when his cash crop is about four to five weeks old, cultivating three times before that to ensure a weed-free seedbed for the cover crop. He has devised an ingenious hand tool for trilling, rather than broadcasting, his cover crop. As illustrated in his book *The New Organic Grower,* a series of three to five single-row vegetable seeders are joined together. With this tool, he has to make only one pass between rows when he is interseeding.

Elizabeth Henderson has been experimenting with cutting furrows for vegetable crops into an established and low-growing cover crop. "The potatoes we grew this way didn't have any Colorado potato beetles," she says.

Bringing New Land into Production

Cover crops and green manures often are used when bringing new land into production. They add organic matter and nutrients, and they also can diminish weed populations. There are three major concerns when bringing land that has been in sod, neglected pasture, weed, or brush into production:

- how to build soil and control weeds in preparation for cropping two to three years down the road;
- what techniques to use if a grower needs to plant a crop the first year;
- how to avoid destroying organic matter or soil structure.

Experienced farmers recommend delaying for at least one year the cropping of land that has been neglected, in order to bring weeds under control and to activate and improve the nutrient status of the soil. Growers who have not taken their own advice admit that planting poorly prepared ground was unprofitable both in yields and labor spent to control weeds. Land set aside for a year or two of improvement, on the other hand, can be made relatively weed-free.

The first step is to take soil tests and to observe the weeds and grasses already growing in a field before making plans or assumptions. A suggested reading on this topic is Ehrenfried Pfeiffer's *The Weeds and What They Tell*, which correlates the plants above ground with the soil structure and nutrients below.[8]

If the soil is naturally fertile and only broadleaf (annual) weeds are present, first-year production may be practical. But if the soil tests poorly or perennial weeds such as bindweed, and grasses such as quack grass, are a problem, delaying planting for a season is the best course. With quack grass, a single plowing and disking actually spreads this pernicious weed, breaking up the rhizomes into many small pieces, each of which can become a vigorous plant.

Some growers recommend a bare-fallow period to control quack, using first a chisel plow and then repeated passes with a springtooth harrow every two to three weeks throughout the growing season to starve the roots and bring them to the surface to dry out. This method is likely to increase erosion and sacrifice organic matter.

Dave Colson says that sorghum-Sudan grass works well to break up old hay sod for vegetable production. In spring, he plows the sod with a moldboard plow, then seeds sorghum-Sudan in June. The sorghum-Sudan gets very tall and thick and outcompetes the summer grasses. In September, he plows the sorghum-Sudan and seeds rye for the winter. Other growers use buckwheat in the same way, often planting two crops back-to-back, or letting the first crop mature, then disking or mowing and allowing it to replant itself. Again, winter rye is planted in fall. Buckwheat seed is cheaper than sorghum-Sudan.

Some perennial weeds and most annuals can be controlled by seeding the land to a year or two of sod, such as a ryegrass/clover mix. Pennsylvania farmer Eric Nordell emphasizes that timely clipping or mowing of these long-term cover crops prevents weeds from setting seeds and starves out unwanted perennials.

Long-term sod can create problems for some first-year crops. For example, wireworms may affect potatoes planted after sod. Tony Potenza says that when he brings a new piece of rented land into production, he plants soybeans if the previous crop was sod. "I may grow soybeans two years in a row. The second year is often better than the first. The sod breaks down over two years. I test the soil the second year and correct for deficiencies then, usually with rock phosphate and liming."

"The value of cover crops is that they're good for the land. They add life," says Jack Lazor. "Once your cycles [rotations] are established, plowing down hayfields can be your cover crop. The only time I cover-crop is on new land. I use a combination of oats and Canadian field peas in a 2-to-1 ratio, and I seed heavily, about 100 to 120 lbs. per acre."

To crop uncultivated land without a year or two of preparation, experienced growers suggest choosing crops that are planted later in the season so there is adequate time for thorough working of the soil. Enough time should be allowed after primary tillage to germinate and kill several generations of weeds before seeding. Large-seeded, fast-growing crops such as corn or beans will compete better against weeds and can be cultivated more aggressively than small-seeded, slow-growing crops such as carrots. Corn can be cultivated repeatedly before and after planting and then overseeded with a weed-suppressing cover crop of ryegrass. Setting out transplants gives a big jump on weeds and does not require a fine seedbed. Crops planted on wide row spacings, such as squash or pumpkins, maximize ease and speed of cultivation. The resulting wide aisles can then be harrowed repeatedly, like a miniature bare fallow, until the vines begin to sprawl.

Perhaps the most innovative and earth-friendly solutions for reclaiming abandoned land utilize animals for brush and weed control. Dick deGraff of Grindstone Farm in Pulaski, New York, has had success using weeder geese to control grasses both in cropped fields and on fallow land. Other farmers report good results using sheep or goats as part of an intensively rotated pasture system to reclaim land grown up to brush. A mechanical alternative suggested by Bill McKentley of St. Lawrence Nurseries involves using a brush blade with large tines that can be adjusted to lift the brush and saplings by the roots so the soil can dry out.[9]

Potential Problems with Cover Crops and Green Manures

Almost all farmers have experienced one or another problem with cover crops and green manures. Timing the plow-down of winter rye is one of the most frequently mentioned. It's important to let any cover crop or green manure decompose for a couple of weeks before replanting the area. This is particularly so with winter rye. This crop has an allelopathic or suppressive effect on many plants while it is first decomposing.

Rye and other grasses also tend to produce a highly carbonaceous residue if left to grow too long into the spring season. This residue can create a rough seedbed and temporarily immobilize available nitrogen while decomposing.

Some growers don't like this practice. "Rye has not worked for us," say Alex Stone and Lance Minor. "It's always too late. It doesn't do anything over the winter, and then it's a weed problem in spring. Undersowing is probably the only way we could incorporate a rye cover into our operation. This requires the right weather conditions," Alex notes.

"When it's sown later than mid-September, rye doesn't put on enough growth to hold the soil. By the time it starts growing in the spring, you need to plow it down for spring crops. But rye can work for growers with more land, who can either plant it in August or plow it down in late spring.

"I'm thinking of doing longer-term sods with compost as part of a better, whole-farm fertility program. But this requires enough land, which not all of us have," adds Alex.

Tough grass stalks also resist being cut, chopped, and incorporated. Many growers note that rye, for example, should be mowed and chopped before it reaches a spring height of 14 to 16 in., and some even report that this strategy reduces its allelopathic effect.[10]

Some farmers are beginning to experiment with winter wheat as a cover crop. "Winter wheat does not come back as persistently as winter rye, and there is no allelopathic effect," according to Will Stevens, a vegetable grower. "Winter wheat grows more slowly than rye and is more leafy. Even though it seems not to give quite as much organic matter as rye, wheat is much easier to plow down."

Other farmers are using an annual or nonhardy plant such as oats or sorghum-Sudan for winter cover. This type of cover, already dead by the time spring comes, readily decomposes once it has been disked and incorporated into the top few inches of the soil. In areas meant for transplants, a dead cover is sometimes left in place to act as a mulch for the first portion of the season.

Cover Crops as Weeds

Preventing a cover crop from becoming a weed generally can be accomplished by tilling it in before seed maturation. With grasses that easily reestablish, some farmers report that the "weed" problem is most serious when soil moisture is high and the crop is fairly young and in its most vigorous growth period.

According to several farmers surveyed by New Alchemy researchers, for example, rye reroots most easily when it's 6 to 8 inches tall. These farmers say that foot-high plants are much less likely to reestablish themselves.

Annual plants often are allowed to blossom before they are tilled in, because they are most vulnerable at this point in their life cycle and their nutrient and organic-matter content is highest. In the case of buckwheat, many farmers also appreciate the bees that are attracted to the flowers. Mowing hairy vetch after it flowers will kill the plant. On droughty soils, allowing plants to bloom can cause problems, researchers note. A crop grown to flowering stage can take up water supplies that the following crop might need. However, none of the farmers involved in this project have reported this effect, probably because northeastern soils tend to be wet in spring and early summer.

Almost all farmers cut or mow their green manures and cover crops several days before they incorporate them into the top few inches of soil. "Let it wilt, let moisture levels decrease, and let it begin to rot and decompose before working it into the soil," they say. This practice avoids putrefaction, or anaerobic decomposition, of excessively succulent plant residues.

Avoiding Tillage

Some growers working small acreages resist cover cropping because they are concerned about the damage tilling does to their soils. Eva Sommaripa says that the deep mulches she uses return as much organic matter to the soil as a cover crop but don't require the extra tillage. "I don't like to till any more than is necessary, but with a cover crop or green manure I don't have a choice. I've used them on new ground—areas where I needed to get rid of the weeds before planting—but I no longer use cover crops on my established beds. I probably would if I were working more land, but I don't need to with this [1-acre] operation. The only thing I might be missing is the nitrogen I could get from a legume crop, so I'm looking for a supply of alfalfa straw for mulch material. I think it could add some nitrogen."

Some growers are experimenting with mowing hairy vetch and then planting through the mat of vegetation without tilling.

David Holm also talks about reducing tillage operations. "I keep about a quarter of my land under cover at all times. On fields where winter rye was planted in fall, I sometimes let it go to seed the next season before flail chopping and disking it in. Then I broadcast vetch over the area, 30 lbs./acre. I like this system because it avoids plowing and saves the cost of new rye seed. However, the succeeding rye crop can come up too thickly. Ideally, I would spread the vetch when the rye was just starting to seed so that there wouldn't be so much rye, but I can't always get to it on time."

David gives another example as well. "I left my brassicas in the field last year for an overwintering site for a parasitic wasp. In early June I disked and seeded buckwheat. It was great forage for bees. After it went to seed, I disked and broadcast rye and vetch at the same time. The buckwheat that came up frost-killed and then the rye and vetch took over. So again, I've avoided plowing that field by just disking."

Insects and Disease

Literature about cover cropping and green manuring sometimes mentions that these crops can harbor pests and disease. Perhaps the best-known instances of this are the attraction of potato leafhoppers and tarnished plant bugs to alfalfa. On July mornings, David Stern and Elizabeth Henderson have noticed clouds of cabbage butterflies rising from a field of red clover and heading for the broccoli. Readers should be aware that farmers in other regions have reported that pests often migrate to other areas when cover crops are mowed or turned under. Consequently, farmers time these operations so that cash crops won't be adversely affected.

Cover crops also can be used to attract beneficials. For example, a healthy abundance of dandelions is an excellent habitat for lady beetles. They will move from their overwintering places under the bark of trees to the dandelions and then onto vegetable crops as the weather warms up.

Endnotes

Chapter 3

1. Anne and Eric Nordell, "A Strategy for Weed-Free Onions," in "Cultivating Questions: Concerning the Bioextensive Market Garden," collected articles from *Small Farmer's Journal* 23, 24, 25.
2. See Steve Gilman, *Organic Soil Fertility and Weed Management*, a handbook in this NOFA series.
3. Vernon Grubinger, *Sustainable Vegetable Production from Startup to Market* (Ithaca, NY: Natural Resource, Agriculture, and Engineering Service, 1999).
4. Grace Gershuny and Joseph Smillie, *The Soul of Soil: A Guide to Ecological Soil Management*, 3rd ed. (Davis, CA: AgAccess, 1995), 38.

Appendix

1. Coleman, Eliot. *The New Organic Grower: A Master's Manual of Tools and Techniques for the Home and Market Gardener,* Revised, Expanded Second Edition. (White River Junction, VT: Chelsea Green Publishing, 1995).
2. For nitrogen values of various cover crops, see the chart on pp. 21–22 of *Managing Cover Crops Profitably* (Burlington, VT: Sustainable Agriculture Publications, 1992).
3. R. E. DeGregorio, Information Sheets released from New Alchemy Institute, May 26, 1986, and June 4, 1986.
4. "Plowdown—A Strategy for the Eighties," *Forage Seed Notes* (Edmonton, Alberta: Alberta Forage Seed Council, 1982).
5. Otto Schmid and Ruedi Klaey, translated by William F. Brinton, Jr., "Green Manuring—Principles and Practice" (Temple ME: Woods End Agricultural Institute, 1981). Publication #2, available from Woods End Agricultural Institute, Orchard Hill Road, Temple, ME 04984.
6. R. J. MacRae and G. R. Mehuys, "The Effect of Green Manuring on the Physical Properties of Temperate-Area Soils," *Advances in Soil Science* 3 (1985); and Mark W. Schonbeck, *Cover Cropping and Green Manuring on Small Farms in New England and New York: An Informal Survey*, (East Falmouth, MA: New Alchemy Institute, 1988), New Alchemy Report #10.
7. Cation-exchange capacity (CEC) is a measure of how well soil and roots can exchange positively charged nutrients (cations) such as calcium, potassium, and magnesium.

8. Ehrenfried Pfeiffer, *The Weeds and What They Tell* (Wyoming, RI: Bio-dynamic Literature, 1981).

9. This discussion of new land is based on a write-up by Eric Nordell and David Stern of an impromptu workshop at the NOFA-NY Blizzard Conference of 1993.

10. Schonbeck, *Cover Cropping and Green Manuring*, 20.

Bibliography

Buchanan, Carol. *Brother Crow, Sister Corn: Traditional American Indian Gardening*. Berkeley, CA: 10 Speed Press, 1997.

Coleman, Eliot. *The New Organic Grower: A Master's Manual of Tools and Techniques for the Home and Market Gardener*, Revised, Expanded Second Edition. White River Junction, VT: Chelsea Green Publishing, 1995.

Gershuny, Grace. *Compost, Vermicompost, and Compost Tea: Feeding the Soil on the Organic Farm*. White River Junction, VT: Chelsea Green Publishing, 2011.

Gershuny, Grace, and Joseph Smillie. *The Soul of Soil: A Guide to Ecological Soil Management*. 3rd ed. Davis, CA: AgAccess, 1995.

Gilman, Steve. *Organic Soil Fertility and Weed Management*. White River Junction, VT: Chelsea Green Publishing Company, 2011.

Grubinger, Vernon P. *Sustainable Vegetable Production from Start-up to Market*. Ithaca, NY: Natural Resource, Agriculture, and Engineering Service, 1999.

Henderson, Elizabeth, and Karl North. *Whole-Farm Planning: Ecological Imperatives, Personal Values, and Economics*. White River Junction, VT: Chelsea Green Publishing Company, 2004.

Managing Cover Crops Profitably. 3rd ed. Beltsville, MD: Sustainable Agriculture Network, 2007.

Maynard, Donald N., and George Hochmuth. *Knott's Handbook for Vegetable Growers*. 4th ed. New York: John Wiley & Sons, 1997.

Nordell, Anne and Eric. "Cultivating Questions: Concerning the Bioextensive Market Garden. Collected articles from *Small Farmer's Journal*, 23, 24, 25.

Parnes, Robert. *Fertile Soil: A Growers Guide to Organic & Inorganic Fertilizers*. Davis, CA: AgAccess, 1990.

Russell, Howard S. *A Long Deep Furrow: Three Centuries of Farming in New England*. Hanover, NH: University Press of New England, 1976.

Sarrantonio, Marianne. *Northeast Cover Crop Handbook*. Emmaus, PA: Rodale Institute, 1994.

Internet Sources

Berton, Valerie. "Diversification in the Field and Market Become Part of Former Conventional Agriculturists' Guiding Principles." SARE Western: Michael & Marie Heath, M&M Heath
http://www.newfarm.org/archive/1000_stories/sare_stories/heath.shtml

Friesen, Mary. Variety and Cooperation the Keys to Pennsylvania Farmers' Success. SARE NORTHEAST: Jim and Moie Crawford
http://www.newfarm.org/archive/1000_stories/sare_stories/crawfords.shtml

Groff, Steve. *No-till Vegetables*
http://www.cedarmeadowfarm.com

Martens, Mary-Howell. Letter from NY: The Art of Crop Rotation
http://www.newfarm.org/columns/Martens/march%202003/index.shtml

Sustainable Pasture Management
http://attra.ncat.org/attra-pub/PDF/sustpast.pdf

UC SAREP Cover Crop Resource Page
http://www.sarep.ucdavis.edu/ccrop/

Index

About the Author and Illustrator

Seth Kroeck has been working and managing organic farms in the Northeast and California for the past eight years. He is currently working his own farm, Crystal Spring Community Farm in Brunswick, Maine, with his wife, Maura, and their son, Griffin.

Jocelyn Langer is an artist, music teacher, and organic gardener, and the illustrator of the NOFA organic farming handbooks. She illustrates and does graphic design work for alternative media and political events as well as organic-farming-related publications. Jocelyn lives in central Massachusetts.

The special farmer-reviewer for this handbook was Eric Nordell; the scientific reviewer was Eric Sideman.

CPSIA information can be obtained
at www.ICGtesting.com
Printed in the USA
FSOW03n1936120217
30645FS